The Site Manager's Bible

The Site Manager's Bible

How to manage any building work, big or small, whether you're a self-builder or home renovator

LEN SALES A.C.I.O.B.

EBURY PRESS

First published in Great Britain in 2006

1 3 5 7 9 10 8 6 4 2

First published by Ebury Publishing
Random House UK Ltd, Random House, 20 Vauxhall Bridge Road, London SW1V 2SA

Random House Australia (Pty) Limited
20 Alfred Street, Milsons Point, Sydney, New South Wales 2061, Australia

Random House New Zealand Limited
18 Poland Road, Glenfield, Auckland 10, New Zealand

Random House South Africa (Pty) Limited
Isle of Houghton, Corner Boundary Road & Carse O'Gowrie, Houghton,
2198, South Africa

Random House UK Limited Reg. No. 954009
www.randomhouse.co.uk

A CIP catalogue record for this book is available from the British Library

Edited by Margaret Gilbey
Designed by seagulls

ISBN: 0091909074
ISBN-13 from January 2007: 9780091909079

Papers used by Ebury are natural, recyclable products
made from wood grown in sustainable forests.

Printed and bound in Singapore by Tien Wah Press.

contents

CONTENTS

2

introduction

As a fourth-generation carpenter, builder and site manager with over 30 years' experience, I am constantly being asked for advice on building work. Since we are now in a boom period for domestic building work and self-build is becoming very popular, I felt that it would be appropriate to write an all-embracing site-management manual that would complement both large and small projects. The main reason for writing this book is to raise awareness of the methods that can be introduced to control and manage a variety of projects.

Although there are many programmes on TV and several good magazines that offer design ideas and outline new products, there is still a lack of quality information to tap into in order to plan and manage a domestic project from start to finish.

I have tried to cover a wide range of the elements that are required to run a successful project, and have given many examples of the types of control methods that are needed. Whatever project you are preparing to undertake, the most important factor throughout will be planning. As with undertaking a journey, if you plan your route, have a good idea of how long it is expected take and what it is going to cost, you should complete your journey successfully.

If I had to prioritise the most important elements of site management, I would list thcm in the following order:

- Planning (generally)
- Fact finding
- Record keeping
- Budgeting
- Specification
- Personnel requirements

- Health & safety
- Information preparation and flow
- Contract of agreement
- Insurance
- Timescales (bar charts)
- Environmental considerations
- Quality control and inspections
- Meetings and communication
- Cost-saving exercises
- Security
- Patience

However, in reality, they are all equally important and their priority will evolve as the project develops.

As I am still actively involved in providing a domestic building service, I take an interest in how other builders and contractors operate. Observation and fact finding prior to selecting your builder or contractor can save you a great deal of time and perhaps disappointment.

If you are planning to employ the services of any building professional, the information and advice I have outlined in this book will help you to put into place crucial procedures that will help to protect your investment. In order to understand these procedures fully, it is vital that you take the time to think properly about the project and what you want to achieve. Only by imparting this information to your hired professionals in a clear and structured manner will you demonstrate your ability as a site manager.

In writing this book, I hope I have provided any aspiring site manager with the necessary information and skills to plan, manage and complete a successful building project. My aim is to provide sufficient knowledge to allow you and all parties involved in your project to carry out the job with complete confidence and peace of mind.

king project file

oject contributors

(main works)

Architect

Quantity surveyor

Engineer

drawings

ocal Authority,

for contractors

ntractors

B
C
D
E
F
G
HI
JK
L
M

Chapter 1

planning and consideration

Benefits of self management

Self management, when done properly, has a number of benefits that this book aims to cover in detail. Firstly, taking the time to read and research specific information is essential. It is sometimes tempting to scan through some sections you feel confident in, and just read the sections in which you may not be so strong. In order to self manage, you need to understand the dynamics of the construction industry and where you can lose a great deal of time and money if your management skills are weak. Taking the time to read all sections in detail may help you to avoid disaster.

The areas that need to be fully understood in order for you to benefit from self management are:

- preparing your files and methods of communication
- providing accurate and detailed information
- maintaining clear and concise documentation and correspondence
- understanding the professionals
- being aware of realistic timescales
- knowing how, and being prepared, to negotiate
- preparing and insisting on legal documentation
- understanding environmental issues
- agreeing on payment procedures
- setting and agreeing standards
- working to a programme and issuing programmes

- obtaining the correct certification and guarantees
- holding progress meetings (formal or informal)
- how to avoid wasting time, material, energy, money
- understanding the importance of health & safety
- keeping your property secure during the project
- how to set up a site: storage, welfare, security, safety
- dealing with demanding situations.

Plan and control

In this book, reference will often be made to 'planning' and 'control'. In an ideal world it would be nice to be able to trust one person to look after your interests and make decisions for you that you would be happy to accept. However, when it comes to decision making on important issues with implications that affect your finances or specification, you need to be in full control at all times.

In order to run any project from start to finish, you will need to understand the importance of planning, and the things you will need to consider. All projects are unique in that there may be subtle differences between your considerations and those of someone carrying out an identical project elsewhere. The first important exercise is to start a **personal project file** or files where you can store and access the information easily. A **working project file** will be explained in Chapter 5.

A personal project file is different from the working project file in that it may contain details of your personal finances and arrangements, and will in any case hold your initial concept details for the project. Therefore you will need to keep this safe and ensure that documents are backed up with copies where required. This file may include the following sections:

List of personal project file contents

These files are fairly self-explanatory, but to understand better how to ensure that they are used efficiently we will break them down section by section. As stated above, it is important to get the planning of any project correct to ensure that mistakes are minimised. You may not require some of the sections shown here; however, this model will give you an understanding of how to put your specific file together.

When you start to gather this information, I suggest you start your personal project file (lever arch) with sections for each of the main elements of the project. Some information you will need to make available (or just have to hand for future use) will start to cross over into other areas and folders. Before long your personal project file will start to resemble a working document that you will be able to retrieve information from easily.

The list of file contents is the very first page that you will see when you open the file and, as in this case, shows sections 1–10: Contact details to General correspondence. You may decide to name them differently or you may have other sections you wish to include.

1. Contact details of all project contributors (prior to commencement)
2. Scheme concept and initial ideas
3. Budgeting details and rough costs
4. Financial provider's correspondence
5. Outline drawings for planning permission
6. Specification of work and building regulations
7. Material quantities and list of suppliers
8. Contractors' details for obtaining quotes
9. Summary of costs and availability
10. General correspondence

1 **Contact details of all project contributors (prior to commencement)**: It is very important to be able to access information easily and at short notice, as the planning stages can be slow. You may need to respond to a letter or message quickly. By having a page that holds information of the specific people involved in your project (listing details such as names, email addresses, direct phone numbers and mobile numbers), you will save a great deal of time and perhaps frustration in sifting through paperwork.

This section is primarily for the people involved in the project prior to the start date; however, there will be some information that will need to be moved to the site file once the project gets under way.

2 **Scheme concept and initial ideas**: This section is useful for storing your initial thoughts and ideas. It is not unusual to refer to them as the project proceeds; it is your initial thoughts and sketches that eventually lead to the project being completed. When you look back and follow the progression of your ideas, it is very rewarding to see them against the end result, and perhaps how different the end result is because of practicalities or advice on design change. As your information for the project increases your initial ideas will eventually be turned into drawings for consideration by the local planning authority/committee; when these drawings are ready they will be stored in the section 'Outline drawings for planning permission'.

At this point you may consider formulating your initial ideas for your specification of work for the architect; he will have a good understanding of what it is that you are looking for. That is not to say your ideas will not change as he may show you alternative materials or present other ideas that you find more attractive. He will, however, need to know the type of materials that you have in mind, in order to provide the correct information on the drawing to meet 'building regulations approval'.

3 **Budgeting details and rough costs**: In order for any project to become a reality, it must be affordable, even if it is a speculative one. The project needs to be financed, and whether you have the funds available, or if you will need to borrow some or all of the money, every facet will need to be broken down and a budget figure set against it. This figure may increase or decrease, and you need to start in this manner so you will know where the money is going to be spent.

By starting off with a list of all of the elements that you need to bring your project together, which you will almost instinctively write down in some sort of sequential order, it won't take long before you realise that your knowledge is starting to grow and, in turn, your confidence and enthusiasm will grow with it.

Always take one step at a time when planning and controlling your project, and be sure that you understand each of the processes before moving on. Most mistakes are made as a result of understanding only half the picture, and if this becomes the norm, the result will be noticed in the financial accounting, and it will also affect how long the project takes to complete.

Self management is very rewarding when done successfully; however, reading this book is only the first step with regard to the research that you need to carry out. You will undoubtedly require some form of professional support. This element can sometimes prove to be expensive, so you must look into what you can do yourself to minimise the cost. In order to achieve savings and to avoid incurring added cost, research and planning is the key. Your time is the main source of this exercise, so starting to plan well in advance of your build date is very important.

Whatever type of project you are planning to undertake, there will be professional services you will need to enlist. It is at this stage that you need to demonstrate that you are in full control, and have put in the effort to make the necessary information available for the professional service providers to do their job. Professional service providers will be discussed in detail in Chapter 2.

You do not need to produce information electronically. However, some people who are proficient in the use of computers may decide to start their own electronic files for correspondence by email, and so on, and so it is important to keep hard copies of all correspondence in the file which can be easily referred to on site and at site meetings. It may also be the case that some of the information that is generated will need to be distributed to third parties, so by providing hard copies there can be no excuse for errors in email correspondence.

An example of this is when you are working out your budgets for specific items, such as roof tiles or ceramic tiling – the information for costing will be used in working out your overall budget, and will therefore become part of your loan application and cashflow forecast, if required, and also for obtaining quotes for the material and labour elements.

On a large project, you may decide to employ a professional quantity surveyor who will work out precise quantities of materials, also taking into account waste and other factors, such as disposing of redundant material and packaging. But later in the book I will be demonstrating how you can work out quantities for yourself, and will give guidance as to what waste factors to apply to the fundamental building processes.

As this book deals primarily with the planning and managing of the building process itself the financing of the project is not covered, other than to say that you need to have a rough figure in mind in order to present your borrowing requirements to your lender.

There are some guides that you can use to establish an initial rate for building whether it be new build, refurbishment, or alterations. Every project is unique and everybody's finishing specification will differ, and so the guides are approximate.

One way of finding out some initial budgeting costs would be to contact a local architect, who will more than likely have had recent experience of a similar project, or may be able to introduce you to someone who has. One thing to take into account when gathering budget costs is that there are many different types of buildings, and the full scope of the work and type of materials needs to be known before costs can be calculated.

For example, particular consideration would need to be given to:

- houses built of stone or materials other than brick or block
- houses with special design features
- houses of a particularly high quality
- houses of larger-than-average size
- properties with more than two storeys, or with basements and cellars
- houses containing hazardous materials, e.g. asbestos
- flats, because types of construction vary considerably, as do responsibilities for common or shared parts
- houses that fall under listed building categories or are in conservation areas and deemed historic. These will almost certainly have to include materials that match the original in quality and age, or at least made to the exact dimensions.

4 Financial provider's correspondence: As previously explained, you should keep files where you can store and access the information for the project easily. It may be advisable to keep this section in a separate file, if you have bank details and personal and sensitive information that you do not want to risk being lost, stolen or viewed by third parties.

5 Outline drawings for planning permission: When your outline drawings have been prepared, you will need to keep copies yourself. It is vitally important that drawings that are revised or have additional notes added are numbered differently with 'Revision A, B, C', etc. It is also important that you mark clearly in red any drawings that may be superseded, including outline drawings. Your architect should issue revised drawings if planning permission has been granted subject to specific changes or requirements. Depending on the size of the project you may need to keep these in a completely separate file due to the amount and physical size of the drawings.

6 Specification of work and building regulations: The specification of work is a detailed description of your requirements and applies to every element of the work. This can be in the form of a plan and/or a descriptive account of the materials, and may include the standard of workmanship that is expected. It is important to pay attention to the spec-

CONSTRUCTION MANAGEMENT SERVICES 99 LTD							DRAWING ISSUE SHEET 1		
PROJECT: EXTENSION - Valkyrie Road, Westcliff-on-Sea, Essex							JOB NO: LJS 666		
DRAWING NO:		ISSUE DATE:					REVISION STATUS:		
		DAY	30	05	06		16	22	24
		MONTH	03	04	04		05	05	05
		YEAR	05	05	05		05	05	05
NO.	DESCRIPTION								
401	EXISTING ELEVATIONS	■							
402	EXISTING GROUND FLOOR	■							
403	EXISTING FIRST FLOOR	■							
404	PROPOSED ELEVATIONS	■					A		
405	PROPOSED GROUND FLOOR	■					A		B
406	PROPOSED FIRST FLOOR	■							
410	PROPOSED FOUNDATION PLAN		■				A		
412	PROPOSED EXTENSION: SECTIONS A-A & B-B		■					A	B
413	PROPOSED EXTENSION: SECTIONS C-C & D-D		■					A	B
420	ROOF LAYOUT			■					
ISSUE TO:		NO. OF COPIES:							
1	LEN SALES	1	1	1					
2	DANE SALES	1	1						
3	MELVIN NAPPER	1	1						
4	CLIENT	2	1	1					
5	ENGINEER	1	1						
6	SAM FOX	1	1	2					
ISSUED BY:									

ification required for the main building materials as some of these will need to meet specific building regulations, and will be referred to on the drawings. We will cover this in detail in later chapters.

Your architect will know what needs to be detailed to ensure building regulations compliance. However, there are alternative materials that can cost less than those stated, and if these are used it is important that you are aware of it and that any appropriate cost adjustments are made.

An example of this is the type of building blocks that will be used for the actual structure of the extension or house. As these large quantities of material will be a major expenditure, you may look at using alternative types of blocks to make a cost saving; however, different makes of blocks can have different properties.

Performance ratings are required in order to satisfy building regulations. Although the blocks are made to an equal size they may not reach specific standards. It is possible, therefore, that if the fixed material is checked by the Building Control Officer and found not to meet the required standard, you may have to rebuild the structure or introduce methods to achieve compliance. Similarly, you may be required to replace the components of a particular piece of equipment that does not comply, such as double-glazed units within the window frames.

Other examples of these materials are:

- bricks
- damp-proof course
- timber
- roof tiles and coverings
- insulation material
- glazing.

Building regulations are a set of minimum requirements designed to secure the health, safety and welfare of people in and around buildings, and to conserve fuel and energy in England and Wales. They are made by the Secretary of State under Powers given by Section 1 of The Building Act 1984.

To gain a clearer understanding of the complexities of these regulations for specific areas, log on to: **www.odpm.gov.uk** which is the website of the **Office of the Deputy Prime Minister**. ODPM's responsibilities for local and regional government, housing, planning, fire, regeneration, social exclusion and neighbourhood renewal put the Office at the heart of the Government's ambition to create sustainable communities for all.

Bear in mind that complying with the building regulations is a separate matter from obtaining planning permission for your work. Similarly, receiving any planning permission that your work may require is not the same as taking action to ensure that it complies with the building regulations.

There is more about the building regulations later in this chapter; but it is important to understand the areas that require compliance and how they are categorised:

The 14 'parts' of schedule 1 to the building regulations are:

A Structure
B Fire safety
C Site preparation and resistance to contaminants and moisture
D Toxic substances
E Resistance to the passage of sound
F Ventilation
G Hygiene
H Drainage and waste disposal
J Combustion appliances and fuel storage systems

K   Protection from falling, collision and impact
L   Conservation of fuel and power
M   Access to and use of buildings
N   Glazing – safety in relation to impact, opening and cleaning
P   Electrical safety

Depending on the type of work that you are undertaking and the values involved you may decide to research some of these specific regulations in detail. The more you understand about what to look for in terms of stamps or certification that indicate compliance, the more you will ensure that you have 'peace of mind'.

Your own personal specification for the finishes both externally and internally need to be well described and, where possible, samples provided in order to ensure that your requirements are being met as well as the building regulations. I will cover the specification and details in other chapters, and some of these will include cost-saving exercises without compromising the required aesthetics.

Alternatively, it would not be unreasonable to ask the builder/contractor to provide you with samples of material that have been detailed in the specification, in order for you to be fully satisfied that they meet your approval, and that they correspond with existing material if required.

You also need to discuss this in detail with your architect, prior to the drawings being prepared. There are many building techniques and materials that meet the same required regulations and standards, some of which may not be acceptable to you. An example of this is timber frame construction with external brick cladding, as opposed to traditional brick and block work. Many people who buy ready-built properties are sometimes disappointed with the type of structure their property is built from, as there are certain advantages and disadvantages with all types.

One example is that timber frame construction is clad internally with plasterboard, and as such will have limitations to the weight of fixtures that can be used, unless the upright timber studs or (legs) of the walls are located. Even if you locate the studs they may not be in the desired position for your required fixing, resulting in a compromise that you are not fully satisfied with. As well as considerations for durability and strength, this type of construction also has different sound performance qualities that may prove to be unsatisfactory for your specific living requirements, e.g. noise transfer.

The traditional method of using solid blocks will eliminate these problems; however, they may add slightly to the construction costs. When the pros and cons are evaluated,

---

I have heard of many people who have bought their dream house on the basis that it fulfils their desired location and specification for size of rooms, etc., only to find out at a later date that the construction is not as they had thought, and they have had to move again in order to obtain their preferred type of construction.

you will be able to make an informed decision on your preferred type of construction. As you can see from this simple example there are aspects that need to be taken into account which you will only understand by making enquiries and carrying out research.

If you have little or no understanding of these types of different construction techniques, it is easy to be disappointed at a later date.

The sample specification (see page 99) illustrates that by writing down your initial ideas you will have to formulate a plan that holds information that will not only help the architect, but will also start to help you break down the costing elements of the project. At this stage you may not know the precise cost of any particular element; however, you will begin to have some idea of where you can make savings by providing some of the material yourself. You will also get a 'feel' for the different contractors you will need to approach for estimates, which in turn will dictate the type and level of information that will be required to obtain detailed estimates or quotes.

This initial specification is the first step to providing a much more detailed specification for the project as a whole, or even room by room. The more detail that you can provide to the contractors for obtaining estimates, the more this will assist in deciding who to use, and where you may be able to trim prices down if required.

7 **Material quantities and list of suppliers:** The better you are able to quantify the materials required, the better it will be to work out your budget cost, and in order for you to do this you will need to understand something about how to measure 'square' and 'cubic' metres. These can be worked out in centimetres or millimetres, and as millimetres are a much more precise way of working out areas, we will be using them for our examples. When tradesmen are working to precise measurements in the construction of particular elements they will use millimetres. It is not advisable to measure in imperial measurement as most material suppliers and contractors will work in metric measurements.

There are 1,000 millimetres to every 100 centimetres. One square metre is a 100cm x 100cm square (about 39.37 inches) and is, for example, the method used for measuring ground floor or wall areas.

Cubic metres are, as the name suggests, an area of a three-dimensional shape in cube form. As above, this is 100cm x 100cm square and 100cm in height, forming a 100cm x 100cm x 100cm cube.

From these examples, you can see that cubic and square areas are much easier to understand from their visual appearance, and are therefore easier areas to calculate. Here is an example of how to work out the area of a wall in square metres by using millimetres as the form of measuring.

If the wall is 2,650mm in height by 4,725mm in length the square metre area would be calculated as 2.650 x 4.725, which equals 12.521 square metres which would be shown as $12.52m^2$.

The square metre (sometimes referred to as the metre squared), is the Standard International (SI) unit of area. The symbol for square metre is m^2. Less formally, square metre is sometimes abbreviated sq m.

The following example shows how to work out the cubic or volume area of a trench that will need to be filled, for example with concrete. By measuring the depth by the width by the length, you will have your cubic metre quantity.

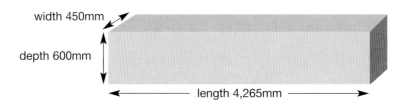

If in this example the measurements were 4,265 long by 600 deep by 450 wide, the cubic metre area would be calculated as 4.265 x .600 x .450, which equals 1.152 cubic metres.

The cubic metre, sometimes called the metre cubed, is the unit of volume in the International System of Units. The symbol for cubic metres is m^3. Less formally, cubic metre is sometimes abbreviated cu m.

By understanding these methods of measuring you will soon be able to work out the quantities of materials that you require, and also provide additional information which will be needed for obtaining your budget costs for material and labour.

Remember that whatever you are taking measurements for, there will always be an element of waste involved, which your material supplier should be able to advise you on.

If you have decided that you are going to have one builder/contractor carry out all of the work on a material and labour basis, you may be of the opinion that he will arrange all the material, etc., and this may well be the case. However, this exercise will give you a better understanding of how costs are accrued. As stated above, you may also need to carry out this exercise for budgeting purposes.

If you are supplying the material you will need to have a good idea of where you will be buying it from; therefore by making a list of suppliers and alternative suppliers for specific materials, you will be able to obtain 'like-for-like' quotes. Always bear in mind that material suppliers may offer discounts as a matter of course, but you will often find that if you push for more you will probably get it.

As far as producing a list of suppliers is concerned, there are the national providers such as Travis Perkins and Jewsons. However, while they are generally good on their main building material prices, you may find that some of the specialised material, such as timber, ceramic tiles and paint, may be cheaper elsewhere, and they may not offer a wide range of these. They will usually provide detailed information on material that is supplied to them by other companies, but by the time they have added their mark up, you may find that you could have negotiated a better deal yourself with another provider so it could be worth shopping around.

Local suppliers will normally provide a good service and may be happy to negotiate costs with you. They may be able to offer products that the national suppliers provide but at a much cheaper rate; however, it is important to ensure that the quality of the material meets your specification. Material such as timber can sometimes be of a poorer quality from local suppliers for various reasons such as limited storage facilities, or the quality of loading and unloading procedures. When material is not stored or stacked in accordance with specific codes of practice, damage or the effects of weather can have a serious effect on its quality. The suppliers may try to pass it on to unsuspecting customers who are not experienced in the quality standards that should be met.

8 **Contractors' details for obtaining quotes:** During the planning of any building project, you will need to start a list of the types of contractors you need to employ. If you leave this consideration to the end you will have lost the opportunity to carry out some research on prospective builders/contractors and may have to make hasty decisions.

> **Many contractors advertise** for work with the advert clearly stating 'no job too small'; similarly you may see them advertise 'no job too big'. This can sometimes give the client false confidence, as most contractors do have their cut-off point with regard to the size of project that they can carry out without overstretching themselves. When contractors do overstretch themselves, they end up providing a limited service to all of their customers. They will sometimes have to relocate their whole workforce to one project in order to satisfy a disgruntled client who has refused to pay up.

If you have decided that you will be using one builder/contractor for the entire project, you will only have to draw up a list of main contractors; but if you are undertaking a project using individual contractors for each trade you will obviously need to draw up a correspondingly higher number of potential contractors to obtain the alternative quotes required. As a benefit of self management, you will be reducing the mark up that main contractors make on the sub-contractors that they use, and this could mean a saving of 10 to 20 per cent or even more. However, this will involve more preparation of individual packages for issuing to individual contractors, and will require detailed specifications for each one of them. To enable you to obtain competitive quotes, you may have to prepare as many as 30 individual packages, depending on the nature of your project.

The different types of packages that you may need to put together to send to specific contractors would include the following elements of construction for an extension:

Groundwork: Site strip, excavation and foundations
Structure: Brickwork and block work, inc. general alterations

External works: Landscaping, driveways, patios
Carpentry: Roof structure, stud partitions, doors, and windows
Roofing: Roof tiling or flat roofing
Electrical: Wiring, data, alarms, etc.
Plumbing: Heating, bathroom sanitary ware, kitchen pipe work
Internal finishes: Plastering, decorating, carpets, ceramics

Specialist suppliers:

- kitchen suppliers and installation
- window suppliers and installation
- air conditioning.

There are many different types of specialist suppliers and contractors and it is important to understand that if the work is of a specialist nature, it is not carried out by the builder.

The detail required for the various types of packages that may need to be prepared for obtaining quotes from different contractors will be covered in Chapter 3.

It is worth checking at this stage whether the builder or specialist contractor belongs to a trade association. If you consider using a company that advertises the fact that they belong to an association, you should ask to see membership details and a copy of their code of conduct.

9 **Summary of costs and availability:** This section of your personal project file is for storing the confirmation of costs that you have accumulated and details of availability of the material. This is vitally important when producing the programme of work.

A couple of things to bear in mind when obtaining quotes for material and labour are that they may have a 'valid until' date. Some material, such as copper or raw materials that could be affected by high demand, may fluctuate. When these materials run low, costs can rise considerably. This can also be said for labour costs; there are many reasons for these to be affected, such as high demand, which will have a marked affect on hourly rates.

10 General correspondence: When you get the project under way on-site you will need to start a site file, or several files on a larger project. At this point it may be useful to have an A–Z section in the back of your file to keep general correspondence. However, if you should find that you are becoming swamped with paperwork and general correspondence, it may be appropriate to start a separate file for this.

Building regulations explained

The building regulations are made under powers provided in the Building Act 1984, and apply in England and Wales. The current edition of the regulations is 'The Building

Regulations 2000' (as amended), and the majority of building projects are required to comply with them. The reason for their existence is to ensure the health & safety of all people in or around the different types of buildings, and this includes domestic, commercial and industrial properties. The regulations also cover energy conservation, and access to and use of buildings.

The building regulations contain sections that deal with definitions and procedures, and what is expected in order to meet the technical performance of building work. The regulations are comprehensive and each section is broken down into categories in order to make them easier to understand. The types of categories include:

- Definition of what types of building, plumbing, and heating projects amount to 'Building Work' and make these subject to control under the building regulations.
- Specification of the types of buildings that are exempt from control under the building regulations.
- Setting out the notification procedures that need to be followed when starting, carrying out, and completing the building work.
- Setting out the specific 'requirements' with which the individual aspects of all building design and construction need to comply, which is in the interests of the health & safety of building users, energy conservation, and of access to and use of buildings.

What you need to do

Anyone wishing to undertake building work which is subject to the building regulations is required by law to make sure it complies with the regulations and to use one of the two types of Building Control Services available. These services are not free and you will be charged for them at set rates which will be proportionate to the size and nature of the work. The two types of services are:

- The Building Control Service provided by your local authority.
- The Building Control Service provided by approved inspectors.

The Building Control Service you select will usually offer advice if requested to clarify specific points before your work starts. Responsibility for achieving compliance with the regulations rests primarily with the person carrying out the building work. If you are carrying out the work personally the responsibility will lie with you. If you are employing a builder or contractor, then the responsibility will lie with them although this can become contentious depending on the terms of your contract with the builder/contractor. It is worth taking the time to clarify this point before work starts as it is the building owner who will be served with an enforcement notice to rectify any work that does not comply with the regulations. This reinforces the importance of using builders or contractors who will work within regulations and ensure that inspections are carried out as and when required.

Under normal circumstances building regulation 'requirements' are covered in the architect's drawings or attached documentation. In order to comply, it is these that you or your builder will need to refer to. If you wish to design and construct your building work in some other way, this is possible provided you can show that it still complies with the relevant requirements that apply to your project. The guidance notes and approved documents, which are to be read alongside each of the 14 parts as specified earlier, **A** – Structure to **P** – Electrical safety, will be taken into account when your Building Control Service is considering whether your plans of proposed work, or work in progress, comply with the particular requirements.

How the building regulations apply in practice

Building work is defined in regulation 3 of the building regulations. The definition means that the following types of project amount to 'Building Work':

- The erection or extension of a building.
- The installation or extension of a service or fitting which is controlled under the regulations.
- An alteration project involving work which will temporarily or permanently affect the ongoing compliance of the building, service or fitting with the requirements relating to structure, fire, or access to and use of buildings.
- The insertion of insulation into a cavity wall.
- The underpinning of the foundations of a building.

If your project is classified as 'Building Work' then the building regulations will probably apply if you want to:

- put up a new building, or extend or alter an existing one, for example by converting a loft space into a living space
- provide services and/or fittings in a building such as:
  - washing and sanitary facilities (e.g. WCs, showers, washbasins, kitchen sinks, etc.)
  - hot water cylinders
  - foul water and rainwater drainage
  - replacement windows
  - fuel-burning appliances of any type.

This means that the work itself will need to meet the relevant technical requirements (specified in Schedule 1 of the building regulations), and they must not make other fabric or services dangerous. For example, the replacement of any double glazing must not worsen compliance in relation to:

- means of escape
- air supply for combustion appliances and their flues
- ventilation for health.

Building regulations may also apply to certain changes of use of an existing building, even if you may feel the work involved in the project does not amount to 'Building

Work'. This is because the change of use to a building may result in it no longer complying with the requirements that will apply to its new type of use. It will need it to be upgraded to meet additional requirements as specified in the regulations, for which building work may also be involved.

Points to consider

The following points should be considered about the site that you propose to build on or make alterations to, particularly in relation to drains, radon and contaminated substances and neighbours.

Drains

If you are considering building or extending over the top of any drainage system, which may be servicing other properties or land drainage, your Building Control Service will need to consult the department that deals with sewerage and associated services. They may make recommendations on what action needs to be taken to protect the drain from any damage which could result from your building work. Many existing rainwater and/or foul water drains are shown on what is called 'the official map of sewers'. This does not mean that if one is not shown on the map there is not one there; therefore if a 'live' drain is found or is known to exist then it is still subject to the consultation procedure. You will need to consider what action may be needed to protect it from your proposed building work.

Radon and contaminated substances

You should also find out if the site on which you want to build has a history of contamination. Many potential building plots may appear to be ideal; however, it is important to rule out any potential problems and to find out if the site may:

- be in an area where the level of radon gas present in the ground is such that there is a possibility that excessive quantities of radon gas could build up in the building
- contain contaminated substances either near the surface or deeper down
- be within 250m of a landfill site.

In cases such as these, protection from gas or contaminated substances may be required under Part C ('Site preparation and resistance to contaminants and moisture') of the building regulations.

Consulting neighbours about your building work

Generally speaking, where planning permission is not required there are no obligations placed upon you to consult your neighbours (other than a moral one), but it would be sensible to do so. The nature of building work and in particular specific problems such as dust and noise could arouse your neighbours' curiosity, so it is advisable to keep them informed about the work for many reasons. You should ensure that your proposed building work does not interfere in any way with their property, as this may lead to bad feeling and possibly civil action for the modification or removal of the work.

Here are two examples of potential problems that can be caused by work that conforms to building regulations but may affect the neighbouring property:

- Work that could result in the obstruction or malfunctioning of your neighbour's boiler flue can be contentious.
- Boundary lines need to be checked to ensure that there are no deeds of covenant which may prevent you carrying out certain types of building work close to or directly adjoining your neighbour's property.

Although consultation with your neighbours is not required under the building regulations as it is with planning permission, you should note that if your project is subject to the Party Wall etc. Act 1996 (which we will cover in Chapter 4) you must give notice to adjoining owners under that Act.

Where building regulations do not apply

You should always be aware of the fact that although the work involved in a building project may not amount to 'Building Work' and consequently not be subject to the building regulations, there are other statutory regulations that it may be subject to. If the proposed project could result in a dangerous situation or damage to your own property or that of your neighbours, it may also result in your own and/or your neighbour's building no longer complying in some way with the building regulations.

Some points to consider here:

- Constructing an open-air swimming pool or a garden pond can cause a danger (particularly to children) during construction and afterwards, and therefore safety precautions will need to be taken.
- Building a garden wall, even a low one, can present a danger if it is not built properly, especially to children, who would not be aware of particular dangers.
- Obstructing ventilation grills to ground floors can cause the malfunctioning of boilers and therefore the flues that discharge the fumes.
- Adding an additional floor covering or decking to an existing balcony may not be subject to control under the Building regulations but may result in a reduction in the effective height of the guard rail, which may increase the risk of people overbalancing and falling.
- The building of an exempt building such as a car port, conservatory or porch which could, for example:
  - obstruct ventilation grills to ground floors
  - obstruct or cause the malfunctioning of wall-mounted boiler flues
  - adversely affect the safety of a gas meter due to reduced ventilation or excessive temperature exposure
  - prejudice safe gas appliance operating conditions.

☻ The removal of a tree close to a wall of your own house or of an adjoining property could affect the foundations and structural stability of the building.

In all such cases it would be advisable to seek professional advice and/or consult your local authority.

Some frequently asked questions

The questions and answers below cover some of the more commonly asked questions about building work and associated services.

Q **If I want to build a new home of any type (i.e. a house, bungalow) or any other type of new building (i.e. maisonettes or a block of flats), will the building regulations apply?**
A Yes – as a new building all of the appropriate requirements in the regulations will apply in full to each element of your project.

Q **If I want to build an extension to my house, will the building regulations apply?**
A Yes – but a porch or conservatory built at ground level and under 30m² in floor area is exempt, provided that the glazing and any fixed electrical installation comply with the applicable requirements of the Building Regulations: Part N 'Glazing – safety in relation to impact, opening and cleaning' and Part P 'Electrical safety'.

Q **If I want to build a garage extension on to my home, will the building regulations apply?**
A Yes – but a carport extension, open on at least two sides and under 30m² in floor area, is exempt, except that any fixed electrical installation must comply with the electrical safety requirements of the Building Regulations: Part P 'Electrical safety'.

Q **If I want to build a detached garage under 30m² in floor area, will the building regulations apply?**
A No – the building will be exempt from the regulations, provided that any fixed electrical installation complies with the electrical safety requirements of the Building Regulations: Part P 'Electrical safety', and it is:

☻ single storey and does not contain any sleeping accommodation
☻ built substantially of non-combustible material
☻ no less than 1m from the boundary of the property.

Q **If I want to carry out a loft conversion to my home, will the building regulations apply?**
A Yes – the appropriate requirements of the regulations will be applied to ensure, for example, that:

☻ the structural strength of the proposed floor is sufficient

- the stability of the structure (including the roof) is not endangered
- there is safe escape from fire
- there are safely designed stairs to the new floor
- reasonable sound insulation exists between the conversion and the rooms below.

You will also need to consider whether your loft conversion is subject to 'The Party Wall etc. Act 1996'.

Q **If I want to convert an integral or attached garage to a dwelling into habitable use, will the building regulations apply?**

A Yes – the appropriate requirements of the regulations will be applied to ensure that the existing accommodation is brought up to the standard required for habitable use, including thermal and sound insulation. All structural alterations to create new window or door openings and the infilling of the existing garage door opening will need to comply with the appropriate requirements of Part A ('Structure'). If the imposed loading is to be increased then the adequacy of the existing foundations will also need investigation.

Q **I would like to carry out internal alterations within my home, or any other type of building – will the building regulations apply?**

A Yes – most likely. The regulations specify the forms of alteration that amount to 'material alterations' and are therefore 'Building Work', taking account of the potential for the proposed work to adversely affect compliance of the building with specific requirements, such as: if your project involves alterations to the structure of the building (e.g. the removal or part removal of a load-bearing wall, joist, beam or chimney-breast); if fire precautions either inside or outside the building may be affected; or if it will affect access to and use of buildings. On the assumption that the regulations do apply, all the work involved in the alteration must comply with all the appropriate requirements.

Q **If I want to replace one or more windows in my home, or any other type of building, will the building regulations apply?**

A1 Yes – if you are replacing the whole of the fixed frame and opening parts. If the work is to your home and you employ a FENSA (Fenestration Self-Assessment Scheme) registered installer, you will not need to involve a Building Control Service. On completion, the installer must give your local authority a certificate that the work complies with Part L and other appropriate parts of the building regulations. You will be provided with a certificate of compliance for your records.

A2 No – if the work only involves general repairs such as:

- replacing broken glass
- replacing damaged or fogged double-glazing units
- repairing or replacing rotten sashes (i.e. opening parts) in the main window frame
- replacing rotten sections of the main frame members.

Q If I want to carry out general repairs to my home, or any other type of building, will the building regulations apply?

A1 No – if the repairs are of a minor nature, for example:

- replacing roofing tiles with the same type and weight of tile
- replacing the felt to a flat roof
- re-pointing brickwork
- replacing floorboards.

A2 Yes – if the repair work is more significant, for example:

- removing a substantial part of a wall and rebuilding it
- underpinning a building
- installing a new flue or flue liner.

> **Where work to roofs** is concerned, if the new roof tiles or roofing material is substantially heavier or lighter than the existing material, then the building regulations may apply. However, if the roof is thatched, or is to be thatched where previously it was not, then the building regulations will apply.

Q If I want to convert my house into flats, will the building regulations apply?

A Yes – the regulations define this as a 'material change of use' and specify the requirements with which, as a result of that change of use, the whole or part of the building must comply with particular parts such as those concerned with:

- escape and other fire precautions
- hygiene
- sound insulation
- energy conservation and contaminants including radon
- the whole or part of the building may therefore need to be upgraded to make it comply with the specified requirements.

Q If I want to convert my home into a shop, will the building regulations apply?

A Yes – the regulations define this as a 'material change of use' and specify the requirements with which the building must comply. The building may therefore need to be upgraded to make it comply with the specified requirements. You should also check with the local fire authority, (usually the Council Authority), to see what 'on-going' fire precautions legislation (such as the Fire Precautions Act 1971) and/or the Fire Precautions (Workplace Regulations 1997) apply when the building is in use.

Q If I want to convert part or all of my shop, office or any other type of non-domestic building into a flat or any other type of home, will the building regulations apply?

A Yes – the regulations define this as a 'material change of use' and specify the requirements which are a result of that change of use. The whole or at least part of the building may need to be upgraded to make it comply with the specified requirements.

Q If I want to underpin all or part of the foundations to my building, will the building regulations apply?

A Yes – the regulations specifically define this as 'Building Work' and therefore the appropriate requirements will be applied to ensure that the underpinning will stabilise the movement of the building. Particular consideration will need to be given to the effect on any sewers and drains near the work.

Q If I want to lay new drains and/or install a septic tank within the boundary of my property, will the building regulations apply?

A Yes – they will apply to new rainwater or foul drains inside as well as outside the building. The building regulations also apply to all non-mains foul sewerage arrangements (i.e. those using septic tanks), including their outlets and drainage fields.

Q If I want to install, replace or alter the position of any type of fuel-burning appliance (including a gas boiler with a flue), will the building regulations apply?

A Yes – The Gas Safety (Installation and Use) Regulations will also apply. However, if you employ a CORGI- (Council for Registered Gas Installers – approved under these regulations) registered installer with the relevant competencies to carry out the work, you will not need to involve a Building Control Service. As there are many types of fuel-burning heating systems it is important to research and use the appropriate installers to ensure that the specific regulations are being adhered to.

Capitalising on your property

Whether you have decided to move to a new address or you are extending outwards or upwards on your current property, it is worth bearing in mind that whatever work you do will add value to your property.

Many people who have lived in a property for a long period of time may not recognise the potential that it has and feel that they will need to move to have bigger rooms or additional rooms. In reality there are very few properties that cannot be extended or altered in order to accommodate additional rooms or, for example, to create a more practical area such as knocking a separate bathroom and toilet into one, or a living and dining room into one.

Most people move to better their current environment whether it is to a larger property, a quieter area, an area more practical for the children to grow up in or a number of other reasons. Unless you have specific reasons for wanting to move, the advantages of extending can be more rewarding when your dreams and aspirations come to fruition. It is worth remembering that when you move there will be costs that can run into thousands when combined, such as moving costs, stamp duty, solicitor's fees, estate agent's fees, time off work, higher council tax, redecorating, and so on.

For those reasons alone, you may well decide to consider investing that money in expanding or altering your current property.

If you have decided that you really do want to move but your current property has not gained sufficient equity, or the area in which you live is noted for properties taking longer than average to sell, it may be worth considering putting together a plan to rent it, in order to cover the mortgage. This may provide you with some profit to put towards another mortgage. Many people have now taken this option, which may be difficult financially, but could in the short term be a way for you to secure a house that you have set your heart on.

Keeping the same address and neighbours

Moving may seem a small price to pay in comparison to having major works carried out on your house. But moving is a big upheaval in many different ways and without doubt one of the most harrowing and stressful events that people experience. Sometimes it seems unavoidable, for instance if you have bad neighbours and have decided to move for this reason – if this is the case there is not much advice that I can offer, except to say that if you find a property that you like, try to gather some information on your prospective immediate neighbours. We have all heard about the 'neighbours from hell', and I am sure that we would all want to avoid that scenario.

On the other hand, there are those who have built up long-standing friendships and are part of a strong and supportive community – this is a consideration that should be thought about carefully before the decision to move is taken.

In the research I carried out for this book I asked the question: 'What do you miss as a result of moving?' The responses were very interesting: high up in the list was having to leave good friends behind and trying to make friends with new neighbours, which in many cases was difficult.

Another question was: 'Subject to gaining planning permission would you have preferred to expand your previous property?' The 'yes' responses were higher than I had imagined, particularly from those who would have been able to develop their original property.

It is true that when we live in a well-balanced and friendly community we assume this to be the case everywhere. However, although it is obvious that we will have to get to know new neighbours, the fact is that we do not really know what to expect. In general, most people are happy to engage in neighbourly conversation. However, it can be very upsetting if you are unfortunate enough to move to a community that does not have the openness you are accustomed to. The more you try to make friends the more frustrated you can become when you do not get the response you would hope for. When this situation arises, the dream house that you always wanted can become more of a place of entrapment, particularly for those who perhaps do not drive, or have particular problems with going out and would normally rely on neighbourly friendships.

It may seem strange to even consider these points; however, times change and with the instability that can exist within some communities now, as opposed to 20 years ago, it needs to be thought about. It is easy to fall in love with a particular property that you see advertised, or when you are driving around looking for the 'FOR SALE' signs. But unless you know the neighbourhood, it is worthwhile carrying out as much research as possible. Moving to a new property can be a very costly and stressful experience, which can be compounded when your expectations are not fulfilled.

> **Unless you are experienced** at moving, and even enjoy it, you may wish to look at the alternatives …

Improve/extend

When we have lived in a property for a long period, we can sometimes fail to recognise the potential that it offers. Firstly, we need to consider what our requirements really are. For example, here are some of the reasons why one may consider carrying out improvements and alterations, as opposed to building an extension – to:

- create a larger kitchen
- enlarge the dining room
- enlarge the living room
- create space for en-suite bathroom to the master bedroom
- create a utility room
- create a play room
- create more storage space.

These are only a few of the factors that could form the basis of your decision making. As the cost of undertaking any kind of building work is considerable in terms of the average income, it is worth taking a little more time to look at the options.

Many properties may offer the space to achieve some of the above without the need to extend. Improvements and alterations may cost considerably less and could be carried out without the necessity of obtaining planning permission, although it is advisable to contact an engineer to ascertain the structural implications when planning to remove dividing walls or chimney breasts, etc. Depending on the age of the property, and other factors associated with, for example, historic buildings, you may need to contact the planning office to ensure that you do not need planning permission to carry out the work.

> **Although reputable builders** will have had experience in carrying out structural work and will be more than happy to offer advice of a structural nature as part of their service, if they are not qualified engineers or do not use the services of a qualified engineer they may well be offering advice that is insufficient. Alternatively, they may offer advice that is over and above what would be required. This in itself may prove to be unnecessary loading!

If you are in two minds about extending or altering a property, it could prove to be beneficial from a financial point of view to start off by making small non-structural alterations. This should in any case add value to a property and make it more desirable to potential buyers. It will also give you the opportunity to see if it meets with your requirements, and if it doesn't you can then plan for the next option.

In general older properties, although well built, were designed with more consideration given to privacy, for example with smaller rooms. The culture of days gone by was one of much more discretion than we are used to today. In addition, older properties tend to have structural qualities that are beneficial to alterations.

That is not to say that all modern properties are lacking in structural qualities; but because of building costs and modern design factors they may require a little more consideration than you would think. For example, many new houses are timber framed, or may have very narrow block walls that do not possess such good structural qualities. And many modern houses tend to have shallow pitch roofs, which would not be at all practical for creating loft conversions or rooms in the roof.

Building a new house

As this book deals mainly with the management of new building work and specific projects to your property, we will keep this section of the chapter brief. However, when you are considering building a new house, it is important to understand that the 'red tape' and 'bureaucracy' are unavoidable.

Deciding on building a new house will require you to carry out much more detailed fact finding and costing exercises as the figures for buying the land and building costs are what most people focus on. It is important to take into account all of the other costs as well, which can be considerable when combined.

Some of these costs can run into many thousands and will depend on factors that may not become apparent until the work is well under way. These may include additional design costs for sub-structure (foundations) if the ground conditions are found to be inadequate for the initial design. This may lead to significant additional costs to achieve the required specification.

An example of this is where a foundation design of a strip footing (trench footing) of 1.2 metres deep becomes a foundation depth of 3.0 metres due to tree roots being

> **In order for you** to understand fully the complexities of house building, it is important that you take time to understand some of the technical requirements and alternatives that may be introduced to make considerable cost savings. I do not cover technical issues in this book; however *The House Builder's Bible, 6th edition*, covers many if not all of the technical and fundamental areas of house building, including costing and environmental considerations.

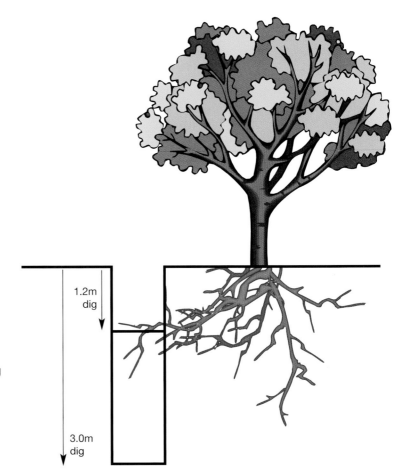

If the drawing shows a
foundation depth of 1.2m
and substantial tree roots
are still visible, the Building
Control Officer would
require the excavation to
continue until the tree
roots are no longer visible

1.2m
dig

3.0m
dig

present in the earth. This is not an uncommon problem and can add considerable costs for the excavation, removal and disposal of the earth and the additional cost of the extra concrete required. This will also have an effect on the timescale of the project, which could then have financial implications with additional interest on borrowed money. For this reason a contingency sum should always be allocated into the build costs, a figure of 10 to 15 per cent would not be unreasonable.

Managing the buying of a plot and the building of a new house from the initial stages will take varying amounts of time to research depending on your knowledge of the procedures, the area, and your resources for obtaining the right information. As detailed in previous sections, the preparation and set-up of your correspondence files will avoid much of the frustrations that can be caused by the amount of paperwork involved. It is important to be able to find any piece of information as and when required in order to maintain continuity.

Local authority planning offices are generally well organised and have their set procedures; however, they can become overloaded with work and this can slow the system down for months on end. It is with this in mind that you need to provide your information and note the dates on which they should be responding to you. Don't be surprised, though, if you do not hear on the day that you expected to. Do chase your information up; but be mindful that, due to the nature of the domestic construction

If you are not experienced in dealing with legal documentation, you should consider placing this element of the proceedings in the correct hands.

industry, and in particular planning approval, etc., there is no obligation for the planners to adhere to specific dates.

If your intentions are to carry out some of the work yourself while managing the project, you must remember that the fewer people there are working on the project, the longer it will take. While this is fairly obvious many self-builders have started off with incredible enthusiasm, only to be worn out and fed up with lack of progress, and eventually end up having to call in the experts. When this situation arises, the incoming experts may take advantage of the situation and increase their prices, as they are aware that when people are desperate they often make the mistake of paying less attention to financial matters.

On the issue of carrying out some of the work yourself, this is of course something that most self-builders will enjoy doing, as it is very rewarding to see your labours resulting in something tangible. In order to manage the work successfully you will need to be on hand to monitor it all on a regular basis to keep an eye on progress and to avoid additional work being carried out, which may have been done without your agreement.

A substantial extension could provide the additional room required for your needs and involves less time, cost and stress than building a new house

Any work that has been designed below ground level is usually based on local conditions, and unless core samples and a thorough survey have been carried out to narrow down the possibilities of additional costs, it is very much an unknown factor. Most cost increases are as a result of additional work below ground, which may include alterations to utility services and drainage.

Core samples and surveys are usually carried out on large projects, where the foundation design and costs can make a difference to the type of structure to be built. A builder may stand to lose money if the contract is a fixed price with no extras, referred to in the industry as 'design and build'.

Such work may include (as suggested above) the additional excavation of footings. This is something that cannot wait for information to flow back and forth, as open trenches are susceptible to collapsing in wet conditions.

It is therefore important to prioritise your time wisely to avoid situations that could result in a costly change in specification due to circumstances beyond the control of engineers, surveyors and architects.

Initial design requirements

Whatever project you are planning to undertake will be based on your actual requirements, whether it is from a practical point of view or strictly investment. It is worth remembering that any changes to specification or drawings may result in additional costs for re-drawing, and could involve additional costs for structural engineer's fees.

With a new build project (whether it is a self-build, major refurbishment or extension), you will most likely have instructed an architect or a specialist designer to produce your drawings. They may assume that you have considered certain aspects of the project and may go ahead with providing the information. You need to consider a few things first, for example:

- Will the design meet our long-term needs (family planning)?
- Can we afford to do more, to increase value later on (profiting from a long-term plan)?
- Will we need to do more later (build in specific elements to make future work easier)?

One of the most frustrating elements of building, both from the builder's point of view and the client's, is having to alter work before it is actually complete just because it hasn't been thought through correctly. To minimise the risk of this happening it is advisable to discuss your plans with as many people as possible, particularly with those who have experience in property development or renovation. By learning from other people's experiences and mistakes, you may save a small fortune in unnecessary work!

⊖ What can we include that we don't need straight away (data cabling, hi-tech cable system)?

These are just some of the considerations, and depending on your circumstances and budget, you may have a list that is fairly comprehensive. Do not try to rush any information through as you may find that the cost implications have a severe effect on your budget.

Self-build

Finding land or a property to renovate is the key to getting a self-build project started. There are many different sources for finding land: there are the usual sources for plots that you would probably be advised to begin with if you know the area where you want to build. Estate agents and auction houses will be able to provide information on available plots, or plots that are soon to be on the market. Local authorities and surveyors may also be able to offer information, and there are many other choices such as the Forestry Commission, church bodies, utility companies and many others.

You could advertise in your local paper or in the paper local to where you are planning to move, and letting estate agents know that you are looking will keep you on top of their priority list. Technology has a major part to play in finding land or property. There are many websites that specialise in this field, for example www.selfbuildit.co.uk

> **From a self-build point of view**, you will need to conduct plenty of research in order to make the savings that can be achieved. Self-build can be both exciting and rewarding; however, one thing is certain, it is challenging even for those with experience. If you have a vision of your dream house you must ensure that the land that you build it on is equal to that vision, or as near as you can get to it. It would be worth the expense of getting an artist to produce an artist's impression of the finished article from a couple of angles – this will certainly give you a better idea of what you will end up with.

There are many self-build houses featured on TV programmes and in magazines that will give you some good ideas to complement your own, and with much reading and discussing the issues you will soon know if the self-build option is for you. As far as your plot is concerned, don't think this will be an easy part of the exercise as these plots are usually expensive.

For accessing specific information and finding out detailed information, The Association of Self Builders (ASB) is a good place to start as they provide support for their members in many areas.

The ASB is able to offer assistance with a wide range of enquiries and has negotiated discounts and special services from a number of suppliers, both nationally and locally (with more in the pipeline), including:

- Jewson's/Grahams; special prices – national discount agreement
- Dulux Decorator Centres; national agreement set up for members to get preferential discounting of high-quality paint products
- First Electrical Wholesale; a huge range of electrical supplies at special prices, i.e. switches, sockets, cable, extractor fans, heaters, etc.
- *Self Build & Design* magazine; 30 per cent off *Self Build & Design* magazine subscription.

Needless to say, they continue to work for their members and will help to negotiate the best deals on all materials such as:

- bricks and blocks
- beam and block flooring
- equipment hire
- welfare facility hire
- site insurance
- boilers and many other types of material.

They hold regular national meetings where you can learn about different aspects of building, meet other self-builders to discuss the real issues of building, managing, dealing with local authorities, and where you will hear talks on techniques and technology. For further information on the ASB, visit their website: www.self-builder.org.uk.

Listening to people who have, in some cases, many years of experience in self-building is the best way to gain first-hand knowledge and advice. Learning how they dealt with situations they encountered during their build may save you a great deal of time, stress and money. The ASB has its own collection of videos covering many aspects of self-build – from foundations to swimming pools.

There is also other useful information available:

- reclaiming VAT
- site organisation
- short bore piling
- contaminated land and the implications for building.

As far as financing a self-build project is concerned there are lenders who deal exclusively with self-build mortgages which will ensure that payments are only made when specific elements of the project are complete. It is worth discussing this with your bank; however, it is a fairly specialised area.

The main difference between a self-build mortgage and a standard house purchase mortgage is that with a self-build mortgage the money is drawn down when specific stages of the project are completed, rather than the traditional method of the whole amount being made available just prior to completion. Some lenders will lend you the money to purchase land, typically 75 per cent of the purchase price or value, whichever is lowest. After the land has been purchased and work is under way, the money for the construction work itself is released in stages. This can be fixed or flexible, depending on the lender. However, there are usually five stages of payment relating to different elements being completed.

During the construction process you can borrow typically 75 per cent of the value of the house as the project progresses, depending on the specific arrangements and the lender's conditions. There are two fundamental methods by which the money can be released during the construction process – these would typically be at the start (advance payments) or end (arrears payments) of each stage. The advance stage payment method enables the self-builder to pay for the work as it proceeds, which will help with cash flow and ensure that bills can be paid when they become due. This advance payment mortgage is very popular; however, in order to achieve this stage, good management is paramount.

With the arrears stage payment method, the money is released after specific stages have been completed and verified by the lender's surveyor, to ensure that the value of work carried out meets the conditions of the mortgage. The difficulty with this method is that it can cause cash flow problems.

The stages of the construction work will depend on the type of building that you have chosen, such as a traditional brick and block house, a timber frame construction, or if you are renovating or converting an existing property.

Here is a typical sample of stages that may be required.

Stage	Brick & Block	Timber Frame	Conversion/Renovation
1	Land Purchase	Land Purchase	Purchase property
2	Site set up and foundations	Site set up and foundations	Structural elements
3	Build to roof level	Timber frame erected and outer skin built	Wind and watertight
4	Roof and watertight	Roof and watertight	M + E and plastering
5	First fix M + E and plastering	First fix M + E and plastering	Second fix
6	Second fix and completion	Second fix and completion	Completion

TYPICAL SAMPLE OF STAGE PAYMENTS MADE BY LENDERS

Insurance considerations for self-build

As with all types of building work, insurance issues need to be considered and with self-build there are many specific elements that come into play. Whereas, for example, an extension to an existing property may offer fairly good security arrangements for equipment and material, a self-build project is a target for theft, and may be seen as a playground for children when in its early stages.

You will require self-build insurance to cover the building of your house, and there will need to be specific elements of the insurance policy that cover other specified elements that the type of project itself will dictate, such as:

- theft
- personal and contractor's accident
- public liability

- machinery
- temporary living accommodation.

As you can see from these examples, the building of your dream house will require much more thought than you may initially have imagined. However, if you approach the right people for advice and act on it, you should not have too many worries with the legalities. As with many specialised fields, there are always people who specialise in certain aspects of each element, so ensure that you approach the right ones and always seek further advice if you are unsure of the advice that you have been given. It does not hurt to have two or three opinions.

Some of the issues that need to be thought about which may affect the type of insurance and level of cover are:

- health & safety
- security
- supply services
- drainage
- environmental issues such as:
  - noise
  - waste
  - traffic movement
  - existing plantation.

Self-build savings (general)

There are definite financial advantages to be made as far as self-build is concerned. It has been reported that savings of between 20 and 30 per cent of the house build price are possible, and in cases where management procedures and competent negotiations have been made, savings of up to 50 per cent are possible.

> **It is a fact** that, under circumstances where the management and research procedures have been comprehensive, the self-build house will be one of a higher specification, and will usually be built with higher-quality materials.

Large property developers who provide development projects of ten or more houses are generally profit driven, however: they tend to use cost-effective building methods that are not conducive to people who are looking for a top-quality finish throughout the whole build. Examples of cost-effective building materials are:

- The type of flooring that is used, such as chipboard interlocking boards that are not easy to lift for future installation of additional cabling.
  - *It would be much more advantageous in the long term to have flooring that can be accessed easily to carry out maintenance and introduce additional services.*

- The use of timber frame housing where the internal walls are made of plaster-board on timber studs. This does meet with building regulations, but it is not very practical when it comes to fixing wall units, etc.
  - *With traditional built block work walls throughout the house, you can rest assured that fixings for shelving and other wall-mounted units will be secure.*

These are just a couple of examples of cheaper building methods that contribute to the speed with which properties are erected. Individuals who research building methods would probably prefer to pay slightly more to have a product that is practical, although it does not mean that all traditional building methods cost more.

On large building development projects, where speed drives the level of workmanship and methods of building, it is not unusual to find that the attention to detail is poor. It is worth taking the time to visit sites where these sorts of projects are being undertaken to see whether the standards that you are setting yourself are being provided elsewhere.

It is worth taking into account that there are property developers who do provide top-quality residential properties of a very high standard, but they tend to be well-established companies who have the financial backing to plan for their development years ahead of the actual build dates. These companies also tend to attract a specific clientele for whom a budget is not high on the agenda.

If you are seriously considering undertaking a self-build project for the first time, do as much research as you can and in as much detail as possible. This will probably be the biggest financial undertaking of your life and if you get it right the experience is a good one. If you get it wrong it will be with you for ever.

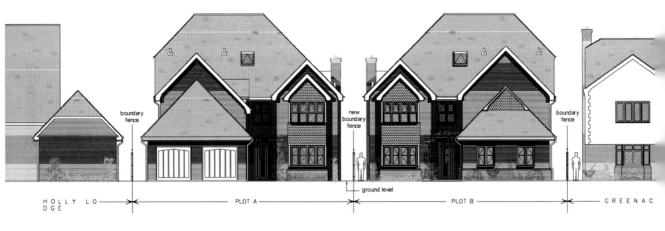

boundary fence

new boundary fence

boundary fence

ground level

HOLLY LODGE   PLOT A   PLOT B   GREENAC

PROPOSED ELEVATION
scale 1:100

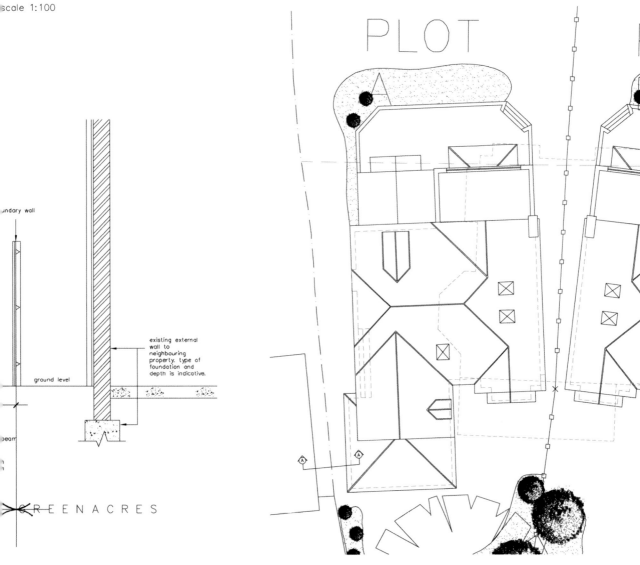

boundary wall

existing external wall to neighbouring property. type of foundation and depth is indicative.

ground level

GREENACRES

SECTION B—B

PROPOSED SITE LAYOUT PLAN
scale 1:100

PLOT

Chapter 2
the nitty-gritty

Professional services

It goes without saying that in order to carry out any type of major building work, there will need to be an element of professional input. The level of this input will obviously depend on the complexity of the project from various points of view. For example, there are now regulations that govern the installation of electrical work and any work that involves installing gas pipes. Some of the professional services that you may need are listed below. It would be very unusual not to come in contact with some of these, even on a small project.

- Architectural Designers
- Architectural Technologists
- Structural Engineers
- Surveyors
- Health & Safety Consultants
- Mechanical and Electrical Engineers

Architectural designers, architectural technologists, structural engineers and mechanical and electrical engineers and surveyors spend many years studying for their profession, and consequently they know exactly what will be required. By using these professionals you can save a lot of time, and they will probably know what the local planning department will be looking for, or will accept. It would be inadvisable to try and produce your own computer-aided drawings for submission to the planning department unless you have good computer skills and a sound knowledge of construction techniques. However, you can still produce plenty of information that an architect will find useful.

Your proposed project and its complexity will dictate how many professional services you will need to employ. If you are using an architect, they will have a database of people who are available for providing specialist information, for example engineer's details and mechanical and electrical specifications. This does not mean that you have to use the specialist designers or engineers that the architect suggests; in fact, by employing them directly you can make considerable savings.

With the age of technology changing at a phenomenal pace it is not unusual to see new programs on the market for producing professional material such as CAD (Computer Aided Drawings). These programs are now available for personal computers and if you have sufficient knowledge of building work in general, and have the time to research information on the Internet, you may be able to produce your own drawings. However, there are many things to consider, as the criteria for meeting building regulations are very complex. If you provide drawings that do not meet the criteria it could have a severe impact on the time it takes for the local planning office to consider and approve your drawings.

Architects

Architects are trained in areas of conceptual design and are therefore able to provide innovative ideas. When there are limitations to what can be put forward for planning permission, they will have a good idea of how to produce the effect you may be looking for.

As the nature of construction is so varied, architects will not generally cover all aspects of architecture, and you may find that some will not touch domestic projects while others may not be interested in commercial projects. It is important that when you are planning to contact an architect, you ascertain what area of architecture they specialise in. Some with large practices may provide a comprehensive range of architectural services, but they tend to be fairly expensive, and their overheads will be reflected in their costs.

Smaller architectural firms that specialise in domestic work may appear to be very cheap and may even advertise the fact that they can provide drawings at a set rate. It is important to remember that in order to obtain a comprehensive quote from a builder they will require as much detail as possible. Where the drawings only show the bare minimum of information, builders may provide a 'guesstimate' rather than a more accurate figure. It is worthwhile taking this into consideration and asking to see drawings that they have provided for projects of a similar nature.

Architectural technologists

Architectural technologists are trained in conceptual design aspects; however, their main field of expertise is that of construction technology. With the pace at which technology is advancing, it is important for architects to be fully aware of the different types of material that are available, for example they may work in conjunction with technologists to achieve new designs and lighter-weight structures.

Structural engineers

Engineer's details will be required for projects that involve structural work of any description, and their role is to come up with solutions that will allow the proposed design to work. In general, where required, the engineer has full responsibility for providing calculations and drawings, which prove that the stability of the foundations and structure will work in conjunction with the design. The information they provide will be included in the information that is eventually sent into the planning office for obtaining full planning permission.

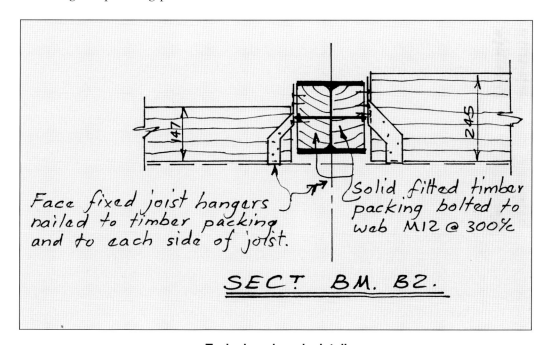

Typical engineer's detail

The planning department will inspect the calculations and drawings in relation to the architectural drawings and any other information that they have on file. They can approve them, but they will request further information if they find anything that they are not happy with.

Surveyors

There are different types of surveyors who carry out various elements of the construction process on both domestic and commercial projects. Below is a list of the different roles that particular surveyors fulfil (surveyors' roles and responsibilities do vary depending on their own levels of experience and training, and the following is a guideline):

Quantity Surveyors:
- calculate quantities of materials from drawings and provide detailed lists
- provide appropriate information for budgeting costs for materials and labour
- provide financial support with regard to valuations and making payments
- arrange lists of materials from specified providers, and negotiate discounts.

Land Surveyors:
- carry out feasibility surveys for developers
- take soil samples for prospective sites
- measure and map sites and boundaries
- provide precise measurements and heights of land
- provide details of local infrastructure.

Building Surveyors:
- conduct condition surveys of buildings including structure, roof, foundations, etc.
- provide reports including defects, cause of defects, advice on remedial work
- inspect work carried out for building regulation compliance.

Property Surveyors:
- conduct condition surveys of buildings for valuations
- provide measurements and descriptive accounts of individual rooms
- provide periodic reports for landlords
- provide maintenance schedules and inspection dates.

As you can see from the above example, there are many elements of surveying that will apply to small as well as large projects, and this is where the surveyor's role overlaps into different areas.

Some surveyors will be experienced and perhaps qualified in all of the above, while others may specialise in just one discipline. If you are going to employ the services of a surveyor, this is where you need to consider your choice carefully.

Health & safety consultant

As health & safety is such a wide-ranging area and one that is fundamental to the satisfactory completion of any project (both in terms of safety and cost), it is important to get it right. If you are employing a builder, the responsibility for health & safety is placed directly on him; however, if you are carrying out some of the work and also employing contractors to conduct some of it, you may need to contact a consultant. Being aware of the risks involved in the project and having a plan to deal with them will help to avoid unnecessary risk-taking by site personnel. If the site personnel are aware that an experienced individual (other than you) will be monitoring and overseeing safety procedures, you should find that they are more co-operative.

Mechanical and electrical engineers

You may have to employ a mechanical and/or electrical engineer to provide schematic drawings and technical specifications if you are planning to employ individual sub-contractors, such as plumbers and electricians. Even so, plumbers and electricians have regulatory bodies which require them to meet certain standards and they must certify the work that they carry out.

The amount of lighting and other electrical equipment installed in the property, including any proposed equipment, must be measured for the amount of electrical current that it will require, and it is the electrical engineer who will provide the necessary information to ensure that your supply feed or anticipated feed is adequate.

Heating and hot water requirements will also need to be calculated to ensure that the boiler size is sufficient to deal with the demands of the property. For example, room size will determine size of radiators, and where there are bathrooms and en-suite bedrooms it is important that the system can cope with maximum demand. It is the mechanical engineer who will provide the information that will detail the equipment and size of pipes that are required.

Professional services' insurance details

Professional consultants who belong to a governing body are required to carry professional indemnity insurance to cover any claims made against them for failures in their services/designs. Although architects, engineers and other technical designers are generally trained to a very high level, occasionally things do go wrong, which may or may not be the fault of the consultant.

For example, if an engineer designed a structural beam intended to be supported on particular bricks or blocks that had specific qualities of strength, which then subsequently was found to be insufficient, the engineer would be liable for the cost of any remedial work. Alternatively, if his calculations were correct and it was the fault of the builder or person responsible for ordering the wrong material, then the engineer would be free from blame.

It is therefore important to ensure that you have a copy of the relevant insurance documents, and that you fully understand them.

Many people are tempted to cut costs by employing the services of non-qualified consultants or 'one man bands'. They may offer very inexpensive rates for a 'full set of working drawings' with planning permission. Don't be tempted to rush into making decisions based on what may appear to be incredible savings on drawings, etc. On many occasions I have seen people who have been tempted by these cowboys, only to have builders and contractors refuse, or be uninterested in pricing the work, due to the poor quality of information.

However, some builders will be happy to provide an estimate for the work based on limited information, knowing full well that they will be entitled to claim for additional work that will clearly have to be carried out. It is likely that if a drawing only holds half of the information for the building work, the estimate will be very loosely put together particularly if there is no detailed specification to support it.

In any of the above examples, if the consultant or builder provides information or work that is sub-standard, it is very likely that if things go wrong it would be difficult or impossible to recover money from them should legal action be necessary. For this reason it is important to insist on obtaining a copy of their professional indemnity insurance, or all-risks insurance in the case of the builder.

Planning permission for domestic properties

The Office of the Deputy Prime Minister provides guides for householders who are planning to extend or carry out work to their home. Copies can be obtained by contacting the ODPM, see details in Further Information at the back of this book.

The following is an explanation of some of the issues that you may come across on a domestic project. Planning permission can become very complicated, depending on the geographical area and nature of the proposal. It would be fair to say, though, that the majority of domestic projects are carried out without encountering too many of the in-depth issues that planning permission can entail.

In general, the planning of any project is a process that will involve third parties and may include objectors. Your understanding of this process may depend on how well your application is dealt with. Even if you consider your project to be small, it does not mean that you will automatically get the application approved. This section should be read in detail as it will help you to provide accurate and precise details, which may be the key factor when the planning committee sit to decide on its outcome. However, in the event that you are not fully satisfied, there is a system of appeal that will be explained later in this chapter.

There are many areas that are covered which are long and sometimes very complex. We will not attempt to look at all of the details of these, but we will provide an insight into what you would need to understand.

Depending on the environment or area in which you live, you may need planning permission or consent for any of the following:

- house extensions and additions including conservatories
- creating a basement
- sun lounges/conservatories
- adding a porch to your house
- swimming pools
- demolition of buildings
- enclosing existing balconies or verandas
- loft conversions
- dormer windows and roof additions
- garages

- garden sheds
- greenhouses
- fences, walls, and gates
- patios, hard standing, paths and driveways
- satellite dishes, television and radio aerials
- decoration, repair and maintenance.

The planning permission system is important for many reasons, both in terms of protecting how sites and communities are developed, and what effects it will have on the environment in general. Planning permission and associated regulations cover many different circumstances, which have very detailed and complex laws. General planning principles, and the procedures for making a planning application, apply equally to owners of houses, freeholders, or leaseholders of flats and maisonettes. There are, however, different rules for flats and maisonettes as they have fewer rights than houses. You can obtain from your local authority the appropriate leaflets for finding out how this affects your particular situation.

Local councils are responsible for planning permission and if you have any issues that need to be discussed or clarified, the first thing to do is to ask the planning department. Your local library may also have information and literature to help in understanding issues that are specific to your local authority. Because there are so many different environmental issues and other factors that need to be taken into account depending on where you live, it is not advisable to rely on general information. The Internet can provide an easy method of finding out what you need to know, though there may be specific local issues that may need to be taken into consideration that only a local planning department would know.

So do you need permission or not?

The following is general guidance about the different kinds of work for which you may need planning permission and those for which you do not. If you are in any doubt about whether you need to apply, you should consult the planning department of your council. They will normally offer advice, but if you want to obtain a formal ruling you can apply for a 'lawful development certificate' by writing to the council with details of the work that you want to carry out. This may require you to pay a small fee, but by obtaining formal correspondence you will start to develop a clear understanding of your rights. This can be particularly useful if you do meet with any objections.

The following are examples of when you may need to apply for planning permission:

- You want to build an extension to your house or garage.
- You want to create a basement to your house.
- You want to build an additional building on your property, e.g. garage.
- You want to make additions or extensions to a flat or maisonette (including those converted from houses).

NEIGHBOURS AND OBJECTORS

Planning permission itself does not allow anyone the right to carry out work on land that is owned by someone else. For example, where planning permission has been granted to your neighbour, despite the fact that you objected to the proposed work or a part of it, you may still be able to take your own legal action to defend any private rights you or your property may have. Needless to say, your neighbours will have the same right as you to object to any proposed development if they are not happy with it. In extreme cases where work has been carried out without planning permission being obtained, property owners have been ordered to put things right later, which is both troublesome and costly. The complete removal of an unauthorised building may even be ordered.

There are many different types of building works and alterations for which you do not need to apply for planning permission. Even if your project is small and you have been formally told that you do not require planning permission, it is worth informing your neighbours about the work that you intend to carry out and discussing it with them. Your consideration will probably ease their concerns that you might be planning work which could affect them. Neglect of good neighbour relations has resulted in many court cases over the years and has often destroyed long-standing friendships. When someone has enjoyed, say, a particular view for years or they have had the benefit of having sunshine until dusk, they will understandably be unhappy at the thought of losing it.

If from a planning point of view you are doing nothing unlawful, there is no point in upsetting neighbours if a compromise could be agreed. It is a fact that people are generally more co-operative with change if they are informed, rather than finding out when the work is under way. Rather than falling out with your neighbour, it may be easier to modify your proposal slightly to take away some of their concerns. If the work that you are proposing to carry out has a serious impact on your neighbours' view from a window or on the amount of daylight they receive, and the window has been there for 20 years or more, you may be affecting a 'right to light' and as such could be legally challenged. In this instance it is advisable to approach a legal consultant who specialises in such cases before you start any work, as you could later be ordered to reinstate existing views or light by removing your development.

If your work does require you to make a planning application, your neighbours and any other third party who may be affected will be given the opportunity to express their views. This is done by letters to immediate neighbours, and a 'yellow notice' (displaying a notice of your intentions on or near the site) being posted on a local street lamp or a post where local residents have the opportunity to see it. If during the work itself you or any of the people working on the project need to enter a neighbour's property, you will need to obtain his or her consent before doing so. For this reason it is common sense to keep on good terms, and by keeping them informed you may avoid any problems.

As everybody's taste varies with regard to the different types of materials used, it is also important to consider that the better the design, the more chance you have of obtaining planning permission. Neighbours can and do have a bearing on whether planning permission is granted, and if your design is not well thought out and presented, you may end up having to spend additional money on re-drawing your plans, re-submission for planning permission, etc., which, of course, would also have an effect on the timescale of the project. It is therefore worth thinking very carefully about how your property will look after the work is finished.

In general terms, extensions and loft conversions will often look much better if the materials used are the same or similar to the existing building. As every project is unique, there is no way to define what is a good design, but when you bought your property, presumably you liked the design, so keeping the same types of features may go some way to appeasing those who may potentially object. Alternatively, if you have a property that has a poor design and your proposals for the new work mean that it will look vastly different from the existing building, it is worth considering additional work to create a completely new look. This may sound contradictory, but in the case of some council estates where the design may be less than flattering, the opportunity to improve the look may be readily accepted by all.

As local councils differ from county to county, you may find that the planning department has design guides or other advisory leaflets which may help you. As previously stated, local designers will have a good understanding of what may be accepted.

- You want to divide off part of your house for use as a separate home (for example, a self-contained flat or bed-sit) or use an existing building, mobile home or caravan in your garden as a separate residence for others.
- You want to build a separate house in your garden.
- You want to build something which goes against the terms of the original planning permission for your house, for example a planning condition may have been imposed to stop you putting up a fence in the front garden because the house is on an 'open plan' estate. You can find out from your local council about specific conditions such as this.
- The work you want to do might obstruct the view of road users.
- The work would involve a new or wider access to a trunk or classified road.
- You want to sub-divide part of your home for business or commercial use (for example, a workshop) or if you want to create a parking place for a commercial vehicle. It is advisable to ascertain if there are restrictions, and indeed if planning permission is actually required.

The above is a general overview of when you may need to apply for planning permission. However, as local councils may differ one from another in their specific regulations, you do need to clarify any proposed development with them.

You do not need to apply for planning permission:

- to carry out internal alterations or work that does not affect the external appearance of the building
- when you want to let one or two of your rooms to lodgers.

Again this is a general overview, and there may be circumstances that require you to obtain planning permission, for example proposed work in a listed building.

Permitted development rights

If you live in a house, you can make certain types of minor changes to your home without needing to apply for planning permission. These rights, called 'permitted development rights' derive from a general planning permission granted not by the local authority, but by Parliament. There are some areas of the country where permitted development rights are more restricted. For example if you live in any of the following areas you will need to apply for planning permission for certain types of work which do not need an application in other areas. There are also different requirements if your house is a listed building.

Areas that require permitted development rights:

- Area of Outstanding Natural Beauty
- Conservation Area
- National Park
- Norfolk or Suffolk Broads.

Your council's powers to withdraw permitted development rights

You should also note that the council may have removed some of your permitted development rights by issuing an *Article 4 direction*. This will mean that you have to submit a planning application for work that would normally not need one. *Article 4 directions* are made when the character of an area of acknowledged importance could be threatened. They are most common in conservation areas. You will probably know if your property is affected by such a direction, but if you are unsure you can check with the council.

Applying for planning permission

If you consider employing the services of an architect or other agent to undertake all of your correspondence regarding obtaining planning permission, you will not need to know the procedure, as they will know what needs to be done and when. However you can make considerable savings by doing this work yourself. Obtaining planning permission can be a long and arduous affair, and when you consider the hourly rate of consultants, depending on the complexity, this could run into several thousands of pounds on large or small projects.

By carrying out this work yourself you will gain a great deal of knowledge about your project, which could be useful for future work on your property or for future projects that you may undertake. Planning officers who work in planning departments are generally very helpful and will help guide you through the procedures. Planning departments have professionals who know and understand the local environment. They will generally be able to give you advice on your proposals, before you have committed large sums of money to having drawings and engineer's details prepared.

The first step

The first step in applying for planning permission is to contact the planning department of your council, inform them of your plans and ask whether they think you will need permission. If they think you do need to apply, they will provide the necessary application forms; the procedures for obtaining planning permission tend to involve a fair amount of form filling. As with all applications of this nature, there are administration fees to be considered, so remember to find out how much these will be, as things may be held up if you fail to include payment with your application.

During these initial stages, it is advisable to find out if they foresee any difficulties that might affect your proposal, and what advice they can offer you in amending it. However, if you have a clear idea of what you want and are prepared to make a strong case to get it, you can submit your proposal as you see fit. But, if your proposal does not reflect what the council would like to see, remember that it could involve additional costs for revising it, and add a great deal of time to your expected start and finish dates.

What type of work are you proposing?

The type of work that you are proposing will dictate the type of application that you need to make, though in most cases this will be a full application. There may be specific circumstances whereby you may want to make an outline application, for example if you want to see what the council thinks of your proposal before you commit yourself to having detailed and costly drawings prepared. You may still be required to submit details at a later stage, but by getting the ball rolling you can start to discuss the details of full working drawings with your architect. You may find that very occasionally, for planning reasons, the council may insist on a full application with all relevant details and drawings even though you would prefer to make an outline application. The council have the last say on matters of this kind as there are likely to be conditions relating to the area or building in question.

Submit your forms

Once you have completed the application forms, they need to be submitted to your local council, together with the correct fee. Each form must be accompanied by the appropriate information showing the work you propose to carry out. The council will advise you what is needed. Adequate plans and drawings are required to be submitted with all planning applications in order for the design to be properly assessed. The information is for the benefit of planners, councillors (on planning and other committees), residents and amenity groups, and may be viewed by others who have responsibility for local issues. They will all assess the impact of the proposed work on the environment and how it fits in aesthetically.

Location plan

A location plan is always required as this will identify the property and any land, together with the boundaries surrounding it. The plan will show where adjoining properties are situated and the local road layout. If the property or site does not adjoin a local highway it will need to show vehicular access roads or tracks.

The scale of the plan will typically need to be in metric and between **1:1250**, and no smaller than **1:2500**.

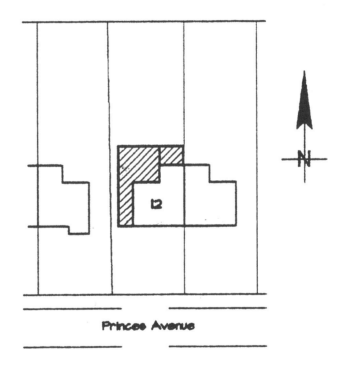

Typical location plan

Details of the existing site layout

The information on this plan will need to show the actual building or buildings that the application refers to, and include any other structures such as garden sheds, workshops, garages or open spaces. In addition, it needs to specify where current car parking spaces are located. You should also consider that where appropriate a tree survey may need to be undertaken, to plot the trees on the drawing. If there are any Tree Preservation

Orders (TPOs), this will require specific attention, which your local planning office will be able to advise you on (see Chapter 4).

The scale of this plan will typically need to be in metric and **1:200**

Details of the proposed site layout

This plan will need to show the position of any new building or extension in relation to the existing building. It will also need to show pedestrian and vehicular access, any changes in ground levels, landscape proposals, including trees and planting to be added or removed, new or altered boundary walls and fences, and any new hard-surfaced open spaces including parking facilities.

The scale of this plan will typically need to be in metric and **1:200**

Other information that needs to be included on the **location plan, existing site layout and proposed site layout** are:

- ◑ north point
- ◑ date
- ◑ drawing number.

Floor plans

In the case of an extension, the drawings need to show the floor layout of the existing building to indicate the relationship between the two, clearly indicating the new work. With alterations, it may be appropriate to combine the layout and floor plan, identifying any demolition that is required.

Include a roof plan where necessary to show a complex roof, or alterations to one.

The scale of this plan will typically need to be in metric and **1:50** or **1:100**

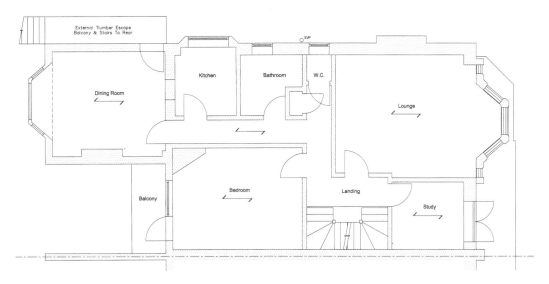

Typical existing site layout plan

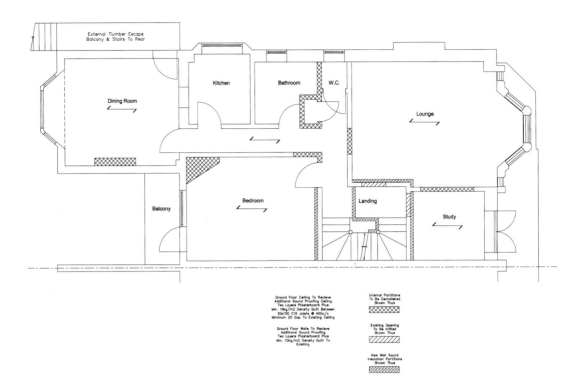

Ground Floor Ceiling To Recieve
Additional Sound Proofing Ceiling
Two Layers Plasterboard Plus
Min. 10kg/m2 Density Quilt Between
50x150 C16 Joists @ 400c/c
Minimum 25 Gap To Existing Ceiling

Ground Floor Walls To Recieve
Additional Sound Proofing
Two Layers Plasterboard Plus
Min. 10kg/m2 Density Quilt To
Existing

Internal Partitions
To Be Demolished
Shown Thus

Existing Opening
To Be Infilled
Shown Thus

New Wall Sound
Insulation Partitions
Shown Thus

Typical demolition plan

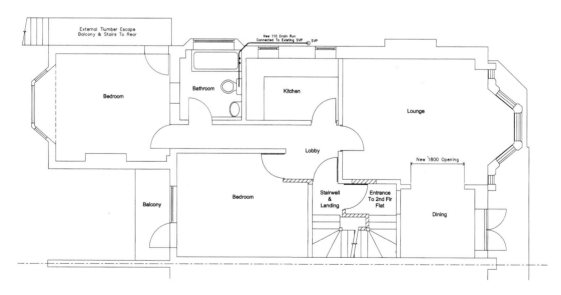

Typical proposed site layout plan

Elevations

All elevations of the property will need to be shown on the drawings, and they must also clearly identify the planned work in relation to the existing structure. The drawing will need to identify the type of material to be used, and it will need to show and describe if this is a change to the existing material, and the type of material that is being proposed. Where appropriate, the external appearance of the building should reflect the changes on the elevation drawings in the context of adjacent buildings. For example, where the property is semi-detached or terraced, the properties either side may need to be included. This would be required to ensure that the proposed work does not detract too much from the features of adjoining properties, but it will obviously be the decision of the planners as to what will be allowed.

The scale of elevation drawings will typically need to be in metric and **1:50** or **1:100** (consistent with floor plans).

Front Elevation
(South)

Side Elevation
(West)

Rear Elevation
(North)

Side Elevation
(East)

Typical elevations

Cross section drawings

Cross section drawings are not always required. However, depending on the nature of the work, it can often be useful to provide them, even if they are not required by the planning department. It really does depend on the complexity of the project as to whether you provide cross section drawings. Builders will be able to use these drawings not only to help them with pricing the work more accurately, but they will also greatly assist in the construction of the work.

The scale of cross section drawings will typically need to be in metric and **1:50** or **1:100** (consistent with floor plans and elevations).

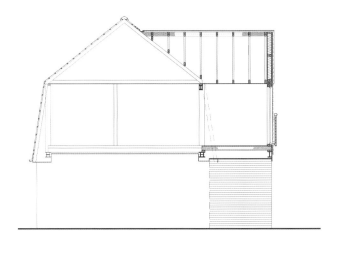

Typical cross sections

Additional information and supporting material

As with any type of proposal that is subject to a planning committee, the more information that you make available may help to speed the process up and clarify any uncertainties that are raised. Types of additional information that you could consider providing include:

- photographs of the property or site and its surroundings
- illustrations, such as perspectives
- artist's impressions
- aerial photographs.

None of the above is actually required for an application, but it may help your cause if, for example, a neighbouring property has a similar design to the one you are proposing, or if your proposal shows that it is not obscuring anyone else's views.

Where aerial photographs are concerned, the expense of providing these could be considerable. On a large project that involves a great deal of input from an architect, it may be worth going to these extremes to show how committed you are to helping the decision process. Showing how the project will fit into the local surroundings may be enough to persuade the planning committee, who under normal circumstances may refuse the application on the grounds of lack of clarity, and perhaps the unknown impact on the local environment.

What the council will do

Planning staff at the council should acknowledge your application within a few days, and normally by post. Your application will be placed on the Planning Register of the council offices so that it can be inspected by any interested member of the public. The procedures for inviting interested parties can include notifying your immediate neighbours, or putting up a notice on or near the site, which is known as the 'yellow notice'.

In certain cases, planning applications may be advertised in the local newspaper, if it is felt that the environmental impact or other considerations need to be brought to the attention of a wider audience. For example, the council may also decide to consult other organisations, such as the highway authority or the parish council, who may be affected by a particular development. The planning department may consider that the preparation of a more detailed report would be required to assess an application fully, which the planning committee (made up of elected councillors) can use to assist in their decision making. The council may give a senior officer in the planning department full responsibility for deciding your application.

You are usually entitled to have a copy of any report submitted to a local government committee, and you are also entitled to see certain background papers used in the preparation of reports. The background papers will generally include the comments (which may be in summary form) of consultants, objectors and supporters, which are relevant to the determination of your application. Such material should normally be made available at least three working days before the committee meeting.

In conservation areas, traditional materials such as Welsh slate
may have to be used, as shown in this example

Planning considerations

The councillors or council officers who have been elected to decide your application must consider whether there are any good planning reasons for refusing planning permission, or for granting permission subject to certain conditions being met. The council will not reject a proposal simply because many people oppose it. They will look at whether your proposal is consistent with the development plan for the area, and this is where your time spent in understanding the planning process will help. Researching and appreciating the sometimes complex issues involved will help you to provide the correct information, and it will also give you a good understanding of why a proposal may be refused.

Other planning issues that the committee may consider include potential traffic problems that may arise from added vehicle movement or parking requirements, or the impact the proposal may have on the appearance of the surrounding area. It may also discuss work that has already been carried out that is similar to your application. You are entitled to inspect the development plan, which will be available by enquiring at the planning department. There may also be associated design guidance for the area in which you live.

How long will the council take?

The council should decide your application within eight weeks; however, if it cannot do so, it will usually seek your written consent to extend the period. Due to the nature of planning and the issues that may need to be assessed, planning can sometimes take longer than the allotted time. If your application has not been decided within this time time, you will be given the opportunity to appeal (see below).

If planning permission is refused or conditions are imposed on the permission (or if the council does not issue a decision), clear reasons must be given for this. If you are unhappy about the reasons or you are unclear about a refusal or the conditions imposed, staff at the planning department will normally be able to expand on any written correspondence. It is disappointing to have a planning application refused, particularly considering the time and effort involved. As the main bulk of the work has been done (e.g. drawings, etc.), changing your plans slightly might make a difference. If your application has been refused, you may be able to submit another application with modified plans free of charge within 12 months of the decision on your first application.

Appeals

If you consider the council's decision is unreasonable, you can appeal to the Secretary of State for Transport, Local Government and the Regions. Appeals must be made within three months of the date of the council's notice of decision. As previously stated, you can also appeal if the council does not issue a decision within eight weeks. Two free booklets, 'Making your planning appeal' and 'Guide to taking part in planning appeals', are available from the Planning Inspectorate, Customer Support Unit, Room 3/15 Eagle Wing, Temple Quay House, 2 The Square, Temple Quay, Bristol BS1 6PN.

Appeals are intended as a last resort and they can take several months to decide. It is much quicker and less stressful to discuss with the council whether changes to your proposal would make it more acceptable.

> **You may have** put in an application for a project that is identical or similar to one that has been carried out adjacent or very close to your property, but this does not mean that yours will be accepted, as regulations and planning conditions may since have changed.

Local authority costs

You will have to pay an initial cost to your local planning department, which would typically be for the administration period up to and including when a decision is made either to grant or refuse your application. At the time of going to press, the cost for an extension to a house or other type of property is approximately £140. For a new property, it is approximately £275.

However, there are other costs that need to be considered, which are payable for building control elements, such as the inspection of drawings and inspection of works. As the type and nature of the work will dictate the level of work that would be required by the building control department, it is not possible to give an accurate or even estimated guide. Your local building control department will have a printed guide that will help you to budget these costs in relation to your specific project. The costs can be payable on a value-of-project basis, or on a fixed-cost basis, which the building control department should be able to quote you. This is worth considering as the value of the project may increase for reasons unrelated to building control inspections or work related to them.

How much are my drawings/professional fees, and so on, likely to cost?

Whatever type of project you are undertaking, the preparation of drawings and other associated professional costs will vary in relation to the size and type of the project. It would be reasonable to expect these costs to be in the region of 5.5 to 15 per cent of the overall building cost, depending on whether your project is a new build or on an existing building. However, they will normally run on a sliding scale and the higher-value projects may bring these percentages down.

Samples of professional fees on an existing building can be seen in the table below.

SAMPLE PROFESSIONAL FEES COST PLAN FOR EXISTING BUILDING								
		Percentages of Fees (approx.)						
	Cost of project	8%	8.5%	9%	10%	11.5%	12.5%	15%
1	Up to £30k							•
2	£30k–£100k						•	
3	£100k–£250k					•		
4	£250k–£500k				•			
5	£500k–£1m			•				
6	£1m–£1.5m		•					
7	£1.5m–£2.0m	•						
8	Over £2.0m	Usually by agreement						

Samples of professional fees on a new build project can be seen in the table below.

SAMPLE PROFESSIONAL FEES PLAN FOR A NEW BUILD							
		Percentages of Fees (approx.)					
	Cost of project	5.5%	6%	7%	8%	9%	10%
1	Up to £10k						•
2	£10k–£30k					•	
3	£30k–£60k				•		
4	£60k–£200k			•			
5	£200k–£1m		•				
6	Over £1m	•					

There are ways in which you can cut down on how much your project will actually cost: the largest cost saving that you can make is on materials that you provide and work you can do yourself. This will entail a great deal of planning on your part; you will need to take a detailed list of the materials required from the drawings and then shop around for the best quotes. This is where your negotiating skills will need to be exercised.

However, it is always worth remembering that every aspect of the work and initial costs are worth discussing, in order to negotiate a better deal. This includes professional fees for people that you employ, but will not include fees that are payable to local authorities. These are set figures that apply to everybody, although, as previously stated, the building control department may quote you a fixed cost based on the work that they envisage carrying out.

Plan your finances

It would be reasonable to expect that anyone considering undertaking a building project of any nature will have thought about the financial implications: how the project will be financed, how much finance is available and what the expected value of the project will be. It is important that these costs are broken down from the start into approximate ('ball park') figures, in order that they can then be refined more accurately as estimates or fixed quotes start to come in. If you are using individual contractors and supplying the materials yourself, this will be a very useful tool. If you are using a builder for most of the project, they will have included most of the building costs for labour and material in their quote, which may include some estimated costs.

You should ask for the estimate/quote to be broken down into separate elements, as it may give you the opportunity to make savings or get reductions. All builders formulate their quotes in different ways and by asking them for a breakdown you will find out if

SAMPLE PROJECT COST PLAN							
Description	Prof.	L.A.	Utility	Mat.	Labour	VAT	Total
1 Plans (Architect/Engineer)	7,000.00					1,225.00	8,225.00
2 Gas, Electric, Water, BT			1,100.00			192.50	1,292.50
3 LA fees		500.00				87.50	587.50
4 Prelims & set-up				8,000.00	1,000.00	1,400.00	10,400.00
5 Site strip				3,000.00	1,200.00	525.00	4,725.00
6 Foundation to DPC				2,500.00	1,800.00	437.50	4,737.50
7 Brick/block to roof				12,000.00	6,000.00	2,100.00	20,100.00
8 Roof & windows				14,000.00	4,000.00	2,450.00	20,450.00
9 Mechanical & electrical				4,000.00	4,000.00	700.00	8,700.00
10 Plastering				1,500.00	3,000.00	262.50	4,762.50
11 Carpentry & fitted elements				3,500.00	3,500.00	612.50	7,612.50
12 Bathrooms				2,500.00	700.00	437.50	3,637.50
13 Kitchen				5,000.00	1,500.00	875.00	7,375.00
14 Decorating & wall tiling				1,600.00	2,000.00	280.00	3,880.00
15 Floor finishes				4,000.00	1,500.00	700.00	6,200.00
16 External works				3,000.00	2,000.00	525.00	5,525.00
17 Contingency				5,000.00	3,000.00	875.00	8,875.00
Totals	7,000.00	500.00	1,100.00	69,600.00	35,200.00	13,685.00	127,085.00

they have indeed worked it out as a proper estimate/quote, or just 'guesstimated'! It is very important to be able to discuss specific elements as there are areas of project specifications that can be scaled down. 'Like-for-like' quotes will be discussed further on Chapter 3.

The foundations, main structure, roof, windows, mechanical and electrical elements are areas where you perhaps will not or cannot make any major savings. However, there will be many other areas where you can, and these you will need to be familiar with.

On the previous page is an example of a typical formatted Project Cost Plan, which enables you to identify specific elements of the project and start to put figures against them.

The following breakdowns of these elements from 1 – *Plans: Architect/Engineer* to 17 – *Contingency* will help you to understand the importance of this level of planning. It is irrelevant whether your project amounts to £20,000 or £500,000: it is probably the largest investment that you are likely to make. By spending time on the planning, you will without doubt minimise the risks of being exploited by the 'cowboy' element that exists in the construction industry, and you will also stand a greater chance of achieving considerably higher savings. In order to assist in the smooth running of the project and in particular the financial element, your plans need to be **SMART**.

- ☻ Specific
  - ▪ The more specific your details, the more likely it is that you will obtain the true cost and therefore present your case for accurate borrowing requirements.
- ☻ Measurable
  - ▪ Having specific details of how the cost of a particular item (labour or material) has been obtained will enable you to measure this (in monetary terms) against alternative suppliers.
- ☻ Achievable
  - ▪ By carrying out this costing process (even with approximate figures), you will soon start to understand the true costs and whether they are achievable with your initial thoughts on budget costs.
- ☻ Realistic
  - ▪ Above all else your earning potential and cash flow needs to meet your planned financial spend to ensure that you can comfortably afford to complete the project.
- ☻ Time bound
  - ▪ All elements of the project should have some sort of timescale, even if specific dates are not known.
  - ▪ You may find that due to the timescales involved in obtaining planning permission, etc., you will have more time to raise additional funds for the project before it starts on-site.
  - ▪ The bar charts referred to in Chapter 4 will assist in producing a timeframe for all elements.

Elements of cost

1. Plans: Architect/Engineer

Architectural and engineering fees will clearly depend on the nature and complexity of your project. However, your architect should be able to give you a fixed cost for his element of the work. The engineer will then be able to provide his costs based on the design features and technical aspects. At this point, he may only be able to provide costs for work needed, so as to supply information that the building control department require, in order to approve the drawings. For a number of reasons, it is fairly common that when the project is under way, the engineer will have to be called upon to revise drawings. For example:

- Unsuitable ground conditions require a re-design of foundations.
- There are insufficient existing foundations for load-bearing walls.
- Walls that were expected to be suitable for loading are found to be crumbling once investigations are carried out.
- There is a change in design as the project progresses.

As you can see, there are many reasons why you may need to ask the engineer for additional calculations and drawings to be incorporated into any revised drawings that the architect may have. This is one area that leads us straight to thinking about the contingency sum. Professional fees are not cheap, and if additional work is required there is generally no compromise. However, an engineer would be unlikely to object to providing small changes or additional information that required only a little of his time to produce. It is advisable to discuss the likelihood of additional costs and at what rate they would be charged.

2. Gas, electric, water, telephone line

The type of project that you are undertaking will depend on whether or not you will need to pay for services that only prescribed contractors can undertake. They need to be treated as separate and identified costs. For example, if the existing incoming gas, electric or water supplies are not sufficient to cope with the additional loadings that will be placed upon them, they may have to be upgraded from the main supplies in the street. You may be required to provide specified size power cables or gas pipes out to street levels. When it comes to the connection in the street, this work can only be carried out by contractors who have experience and knowledge of where the services are located, and who are certified to carry out the work.

3. Local authority fees

As previously mentioned, the actual cost will depend on the type, nature and complexity of the project. In general, these costs should be small in comparison with others. However, you will not be able to make significant savings here, as administration of planning permission and building control procedures, and so on, are usually fixed.

In the event that you need to arrange a road closure or have scaffolding on the public highway, there may be additional costs. There may also be a deposit required, which in some cases can run into thousands of pounds. This is because the local authority puts control measures in place to avoid taking individuals or companies to court for work that has had to be carried out, due to a building project causing damage to, for example, a path or highway.

4. Preliminary costs and set-up

As with any project, there are always preliminary costs that will need to be allowed for in order for the project to operate. These costs can generally be combined under one sum and then broken down into individual elements that can be identified, for example:

- **Scaffolding**
  - When obtaining a quote for scaffolding, always allow 10 per cent towards your contingency sum as it may need to be adapted as the project progresses.
- **Skips or 'muck away' lorries**
  - It is likely that if you are undertaking a project that involves large quantities of spoil to be removed, a 'muck away' allowance will be made as this is a cheaper option than using skips.
  - However you may still require a sum for skips, as this is the best method for rubbish removal during the later stages of the project when large lorry movements may be less practical.
- **Temporary power and water supplies**
  - Apart from the overall cost of your electrical and plumbing work for the project, there may be a requirement for temporary supplies, which may need to be adapted or moved as the project progresses.
  - This work will obviously need to be carried out by experienced personnel, and the correct material used to avoid placing anyone in danger.
  - This material may become redundant after use and may be more expensive than you anticipated.
- **Hoarding**
  - In order to make the site secure and safe you may need to erect a hoarding around it, which can prove to be expensive on a large plot.
  - There are obvious alternatives, such as hiring lightweight wire galvanised fencing panels, but they are not as secure as a timber hoarding.
  - If you have material and hired plant on the project, you need to think about making the site as secure as possible.
- **Plant hire (mechanical)**
  - This is an area that needs some serious thought. It is generally considered that hiring small dumpers or any mechanical plant is expensive, However, when there is a great deal of earth or debris to be moved they can pay for themselves in a short space of time.
  - Take the time of physical labour for moving debris to a skip (including the loading and unloading element) and you will soon see how cost savings can be made.

- Security of the equipment is one of the major issues to take into account. Small mechanical elements of plant are easily moved and therefore may be easily stolen.

◉ **Small tool and equipment hire (electrical)**
  - It is inevitable that you will need to hire small tools or equipment, for example an electrical breaker (Kango) or water pump.
  - You may need to hire a tower scaffold at some stage, which may be a more practical solution to providing a rigid scaffold or adapting an existing scaffold.
  - It is advisable to obtain a catalogue with the cost of hire equipment and then open an account with a negotiated discount rate. It may be possible to secure a discount rate of 30 per cent or more.

◉ **Health & safety measures**
  - There are many health & safety issues to take into account that affect not only the site activities, but also the local environment. These will be covered in Chapter 6.
  - Types of material that need to be provided as part of the site 'prelims' are first aid equipment, and personal protective equipment, such as hard hats, boots, gloves, eye protection, ear defenders, etc.
  - Depending on the type and details of the project, you should be able to identify the necessary equipment required to protect the general public.

◉ **Temporary office space and welfare facilities**
  - If your project is of a large nature you may decide that an office set-up would be beneficial to the running and security.
  - Toilet facilities will be required for any project and it is important that sufficient facilities are provided for the number of personnel on-site.

◉ **Telephone service**
  - The cost of a telephone needs to be calculated into the 'prelims' and depending on the timescales involved it would be worth having a main line installed rather than relying on a mobile phone. The number of essential calls will be enormous, even for small projects.

◉ **Secure storage container**
  - A small or large container may be hired to store general materials, mechanical plant and tools.

◉ **Insurance**
  - Although we will discuss insurance in Chapter 4, the cost needs to be allowed for in your 'prelims'. Depending on the nature and type of project, you may need a completely separate insurance policy.

5. Site strip

When starting any project, it is possible to identify some costs that are not likely to change. The site strip is one where all elements can easily be costed. It is a process that allows the top layer of earth or material to be removed to a depth where the foundations can be marked out before excavation takes place.

There are circumstances that may affect the site strip, such as extreme weather conditions, causing a serious 'mud issue'. Where this is a problem you may find that you are responsible for cleaning up the highway where the lorries have left a trail of sludge or mud. When this happens there are other considerations such as where the sludge will be moved to. If you just wash it down the drain and the drain subsequently blocks up, you will be responsible for the cost of removing the blockage.

You need to allow a contingency for this or at least plan how you will deal with a situation of this nature both in the site strip and excavation of foundations. There are road cleaners that can be hired, or wheel wash systems that can be introduced to the site where the lorries' tyres are cleaned before they leave site.

Elements of site strip that can be identified may include:

- **Hire of digger and driver**
  - Groundwork contractors can make huge profits on this type of work. It is worth weighing up your options of either using a contractor to carry out all of the groundwork operations, or break the elements down and organise them yourself.
- **'Muck away' lorries based on amount of spoil to be removed**
  - Bear in mind that as the spoil is excavated, it will 'bulk out'. This means that as the earth/spoil is broken up by the mere fact that it is being removed in relatively small amounts, it will increase in volume as it becomes aerated. The reason for this is that in its natural form in the ground it is one solid mass with no pockets of air, but when it is broken down into smaller amounts there will effectively be pockets of air that are trapped in the mass. This can increase the mass from its original measured area by as much as 35 per cent, depending on the type of spoil.
- **Road cleaning or wheel wash system**
  - As described above, this is an area worth considering for a contingency sum. Depending on the project, it may run into hundreds or thousands of pounds.

6. Foundation to damp-proof course (DPC)

This can involve many different elements and only the project specifics will dictate what is involved. However, some of these elements could include:

- excavation including machinery and removal of spoil
- drainage and manholes
- shuttering where ground conditions are too soft
- reinforcing bars
- concrete to excavation
- brickwork to DPC

- hardcore and blinding (sand)
- plastic membrane
- concrete to over-site.

7. Brickwork and block work to roof

This element will be one of the most expensive as it involves the construction of both the inner and outer skin and includes the forming of the first floor. This will need to be incorporated as the walls are constructed in order to provide the necessary strength for the building. The breakdown of this element would include:

- brick and block work including all associated material
- lintels
- floor joists (first floor) and any additional floors
- cavity insulation and wall ties
- cavity closers at window and door jambs.

8. Roof and windows

This will also be a major cost and would generally include the following in order to make the building secure and weathertight:

- all roof trusses or timber for forming the roof
- roof felt or breather membrane
- roofing battens
- roof tiles, vents, etc.
- lead sheeting
- fascia and soffit
- windows and external doors (including glass).

9. Mechanical and electrical (M + E)

Once the heavy elements are complete and the building is secure and weatherproof, the plumbing and electrical elements can start. This work can be started beforehand where there is no danger of materials falling from above. As these elements are now regulated by building control requirements, and must be carried out by suitably qualified personnel, it is likely that they will form separate packages whereby the contractor would normally supply the majority of the material. However, there are elements that can be taken out of their package which may form part of your cost savings. The elements in this section could be broken down as follows:

- **Mechanical contractor (plumbing and heating) package could include:**
  - providing all temporary services
  - supplying labour and material for installing all heating and gas pipes, hot and cold water feeds, including lagging
  - connection of all sanitary ware and kitchen equipment and appliances, including running waste pipes
  - fitting all rainwater goods, guttering and down pipes.

- Electrical contractor package could include:
  - providing all temporary services
  - supplying all cables and associated electrical equipment to provide main supply up to sockets, switches and lights
  - supplying all other cables for telephone, TV, data cabling, etc.
  - cabling for alarm system.
- It is worth noting here that builders/contractors will have a mark-up for materials which may form part of their profit margins. It is also worth considering providing some elements yourself, particularly where you can negotiate discounts with suppliers. Materials that you could provide include:
  - boilers
  - sanitary ware
  - radiators
  - kitchen equipment, sinks, bowls, etc.
  - shower cubicle
  - rainwater goods
  - wall socket face plates
  - light switches
  - lights, internal and external
  - alarm equipment.

Mechanical and electrical (M + E) drawings may be required on larger projects, showing pipe and cable routes. The drawing would also detail the technical aspects that may be required and could typically include:

- Mechanical
  - Pipe diameter
  - Hot or cold pipes
  - Valve positions
  - Direction of flow
  - Lagging requirements
  - Pipe drops
  - Position for risers (service ducts)
- Electrical
  - Meter head position
  - Distribution board position
  - Armoured cable routes
  - Cable routes and type
  - Socket position of cable
  - Switch position
  - Light positions
  - Service ducts
  - Earthing points

Coloured leaded light glass can make a lovely feature in a period home

Where there is no need for specific M + E drawings the information required for determining the position of sockets, lights, switches, radiators and all other associated equipment should be clearly marked on the main drawings. If you do decide to have M + E drawings produced, the cost is likely to run into hundreds of pounds. However, it is worth the investment on projects where there are either fundamental changes to existing systems, the size of the project is such that drawings are required, or the technical details are significant.

10. Plastering

This is probably one of the easiest elements to provide a figure for. It involves measuring the wall area for all walls excluding windows and doors, and then working out a square metre rate for the type of plastering to be carried out. There are different rates for plastering with sand and cement as opposed to gypsum plaster, and there are different rates again for plastering on plasterboard. There are other alternatives such as dry lining but, in general, wall areas do not change. Coving is normally carried out by the plasterer and would therefore be calculated in this element.

> ◐ Material content of plastering could include:
> - sand (loose or bagged)
> - cement (bagged)
> - plaster (bagged)
> - angle beads and stop beads
> - coving and adhesive.

11. Carpentry and fitted elements

It goes without saying that the carpentry element can be a huge part of the costs. However, for the purposes of this costing exercise we will cover the basics, also pointing out elements of carpentry that you may wish to carry out, but due to budget constraints may have to be left until a later date.

● In order to complete the project with the basic carpentry elements in place, you would need to consider the following:
  ■ stairs, handrail and spindles
  ■ floorboards
  ■ studwork or partition walls
  ■ plaster boarding
  ■ door liners
  ■ doors and door stops
  ■ architraves
  ■ skirting
  ■ door furniture
  ■ window boards
  ■ curtain battens
  ■ loft hatch
  ■ bath panel.

So these are the basics that would need to be in place in order to complete the carpentry package. They may be the minimum requirement that a lender may look for in order to meet the stage payment criteria.

● Other elements of carpentry that could possibly form part of your specification classed as fitted elements might include:
  ■ pipe boxing (waste pipes)
  ■ shelving (airing cupboard)
  ■ boarding to loft area (storage)
  ■ fitted wardrobes
  ■ radiator covers
  ■ fire surrounds.

12. Bathrooms

Bathrooms and kitchens are known to have a major influence on house-buying decisions, and so for the bathroom it is worth considering buying good-quality sanitary ware. If you have decided to employ plumbers to carry out the labour element of plumbing and associated works (with them supplying the pipe work and fittings), you need to research sanitary ware suppliers as there are huge cost differences between the various companies providing this equipment.

- Materials may include the following:
  - baths
  - toilets
  - bidets
  - shower bases and cubicles
  - shower controls
  - taps and fittings
  - towel radiators.

13. Kitchen

As stated above, kitchens are of key importance when selling a property, so it is worth buying good-quality units and fittings. This element is one where extensive savings can be made depending on your preferences and, of course, budget. It is possible, for a normal-sized family home, to spend anything from approximately £1,500 up to £15,000 on a kitchen. You will need to spend some time looking at different examples in order to weigh up your options. And it is important that the kitchen is in keeping with the type of property you have.

- Additional items that may be required when buying a kitchen and associated equipment include:
  - base and wall units (pre-assembled or flat pack)
  - doors and drawers (various options)
  - door and drawer handles
  - worktops (various options)
  - cornice and pelmet rail.

At this point, it is worth considering whether or not you will be buying self-standing appliances, or if you will be having integrated appliances that are built in with the units and concealed behind the doors. Integrated appliances will look much better since they work in conjunction with the opening and closing of the doors. However, if at a later date you decide to move to a property that did not have these, you might have to buy them all over again, as most purchasers would expect these appliances to be left in the house as part of the fixtures and fittings.

- Typical integrated appliances could include:
  - cooker and hob
  - fridge
  - freezer
  - washing machine
  - dishwasher
  - tumble dryer
  - microwave
  - extractor unit.

As you can see from the above, replacing all of the appliances that you leave behind could amount to a considerable sum. However, spending more on the kitchen in the first place could add a substantial amount to the selling price. Modern kitchens and expectations of house buyers are geared towards integrated systems, and in the event that you did move, you would more than likely look for a property that had similar qualities.

14. Decorating and wall tiling

It would be fair to assume that if you were building an extension or loft conversion, you would decorate the new work to suit the existing finishes, unless you decided to completely redecorate. There are many types of decoration material and qualities of paper, for example, so you would need to draw up your specification in some detail, room by room. Once you have your specification, you can break this down in much the same way as for the plastering. Wall and ceiling areas are easy to measure and will determine the amount of paint or paper that you require. Likewise with wall tiling; this is an area where you would not expect to have much waste as tiled areas can be exactly defined.

The following room schedule will help to break down the quantities of material and will help to allocate a cost to each room. The information on this schedule would be taken from a written specification for each room. By providing this with the amounts blanked out, individual decorating contractors would be able to supply a breakdown of their estimate/quote.

	SAMPLE COSTING SCHEDULE FOR DECORATING AND PANELLING TO WALLS AND CEILINGS								
	ROOM	**Paint**	**Paper**	**Panelling**	**Tiling**	**Mat.**	**Lab. no VAT**	**VAT**	**Total**
1	Hallway/stairs/landing	Yes		Yes		150.00	200.00	26.25	376.25
2	Living room		Yes			130.00	250.00	22.75	402.75
3	Dining room		Yes			125.00	220.00	21.88	366.88
4	Kitchen	Yes			Yes	150.00	300.00	26.25	476.25
5	Study			Yes		300.00	600.00	52.50	952.50
6	Downstairs w/c	Yes		Yes	Yes	120.00	250.00	21.00	391.00
7	Playroom	Yes				70.00	120.00	12.25	202.25
8	Conservatory	Yes				40.00	50.00	7.00	97.00
9	Bedroom 1		Yes			200.00	250.00	35.00	485.00
10	Bedroom 2		Yes			175.00	200.00	30.63	405.63
11	Bedroom 3	Yes				100.00	100.00	17.50	217.50
12	Bedroom 4	Yes				80.00	100.00	14.00	194.00
13	En-suite Bedroom 1	Yes			Yes	315.00	320.00	55.13	690.13
14	En-suite Bedroom 2	Yes			Yes	220.00	260.00	38.50	518.50
15	Family bathroom w/c	Yes			Yes	400.00	370.00	70.00	840.00
16	Shower room	Yes			Yes	350.00	300.00	61.25	711.25
17	All ceilings	Yes				250.00	500.00	43.75	793.75
18	Storage cupboard	Yes				30.00	40.00	5.25	75.25
	Totals					**3,205.00**	**4,430.00**	**560.88**	**8,195.88**

In many ways these schedules help to provide the necessary information at a glance. For instance, you could use one for your room-by-room specification.

SAMPLE ROOM FINISHING SCHEDULE											
Bedroom 1	**Paint**	**Stain**	**Matt**	**Gloss**	**Eggshell**	**Tile**	**Paper**	**Lam.**	**Carpet**	**Brass**	**Chrome**
1 Skirting	Yes				Yes						
2 Architrave	Yes				Yes						
3 Wall finish							Yes				
4 Floor finish								Yes			
5 Ceiling finish			Yes				Yes				
6 Doors		Yes									
7 Door furniture											Yes
8 Switches etc.											Yes

15. Floor finishes

As you can see from the schedules, they have an important part to play in narrowing down your costs, together with the material and labour content. It does not matter about the size of project nor at which point you are planning to carry out any finishes: you will still need to know what these will be and the schedule will help to identify them.

SAMPLE FLOOR FINISHING SCHEDULE								
ROOM	**Laminate**	**Carpet**	**Tiles**	**Vinyl**	**Mat.**	**Lab. no VAT**	**VAT**	**Total**
1 Hallway/stairs/landing		Yes			300.00	115.00	52.50	467.50
2 Living room		Yes			400.00	65.00	70.00	535.00
3 Dining room		Yes			370.00	50.00	64.75	484.75
4 Kitchen				Yes	120.00	40.00	21.00	181.00
5 Study		Yes			250.00	70.00	43.75	363.75
6 Downstairs w/c				Yes	30.00	20.00	5.25	55.25
7 Playroom	Yes				200.00	120.00	35.00	355.00
8 Conservatory	Yes				320.00	150.00	56.00	526.00
9 Bedroom 1	Yes				250.00	175.00	43.75	468.75
10 Bedroom 2	Yes				180.00	145.00	31.50	356.50
11 Bedroom 3		Yes			275.00	55.00	48.13	378.13
12 Bedroom 4		Yes			175.00	45.00	30.63	250.63
13 En-suite Bedroom 1			Yes		460.00	155.00	80.50	695.50
14 En-suite Bedroom 2			Yes		320.00	140.00	56.00	516.00
15 Family bathroom w/c	Yes			Yes	125.00	50.00	21.88	196.88
16 Shower room			Yes		165.00	75.00	28.88	268.88
17 Storage cupboard	Yes				60.00	30.00	10.50	100.50
Totals					**4,000.00**	**1,500.00**	**700.00**	**6,200.00**

> **If you have newly built** and plastered walls, it is advisable only to apply a coat of paint in order for the structure to dry out and settle. The heat within the property will gradually dry not only the walls, but all materials that have been used, and it may be weeks or months before settlement cracks stop appearing. Don't be alarmed by these cracks: it is a natural occurrence and one that is easily remedied. You will need to keep some of the same colour paint so that you can make good these cracks. If you hang wallpaper on walls that have not dried out sufficiently, the chances are that it will tear as the cracks appear.

16. External works

As external works are usually one of the last elements of the project to be undertaken, it may be worth having a general plan and desired layout that can be graded up or down depending on how the cost of the overall project works out.

17. Contingency

There are many reasons for including a contingency figure, and it is not unusual for the whole amount to be used. Additional professional costs may be required for re-drawing or for additional work by the engineer. Work in the ground (e.g. foundations) can sometimes bring unexpected expense, such as the excavation needing to go much deeper than planned. In events such as this, associated costs are going to be much higher.

Although these are only initial thoughts on costs, you can now start to understand how the time and effort put into this exercise will help you to narrow the costs down, so that you don't end up with too many unpleasant surprises.

Credit from suppliers

Credit from suppliers is a very important part of all industries with regard to cash flow and can be an important factor when planning how your finances will work. You may be able to show the individual suppliers on a business plan or general expenditure plan for your project when applying for a loan. Builders and contractors will usually have accounts with one or more of the national building material suppliers and many more local suppliers. While the national suppliers deal mainly with builders and sub-contractors, they now have special arrangements whereby the general public can set up accounts and receive the same terms and conditions as builders.

With self-build packages and projects that are financed by banks or mortgage providers, some builders' merchants have a specific plan for supplying material that is linked to the programme of the project and in particular when payments from a lender will be made for the work.

Credit from contractors

Builders and contractors will normally expect to wait until they have carried out all or some of the work before getting paid; this, of course, will depend on the size of the project. When it comes to your cash flow it is important that you use to your advantage all of the resources available to you for financing your project. This is particularly relevant where stage payments are being made by a lender.

As with using credit facilities or arrangements that you have made with material suppliers, your chosen builder may be prepared to enter into a similar arrangement that is dictated by your lender. Don't forget that the builder will also have credit facilities for material that he is supplying, and in effect you could be paying for the work before he has paid for the material!

It is slightly different when it comes to labour-only contractors. Unless specifically run as a company, these work for an hourly, daily or weekly wage. Although it would not be unreasonable to pay labour-only contractors one week in arrears, they may expect to be paid at the end of each week for the work that they have carried out. This is a point that needs to be discussed with them. However, from a cash flow and quality control point of view, it is advisable that you pay one week in arrears and to state this in your terms of agreement.

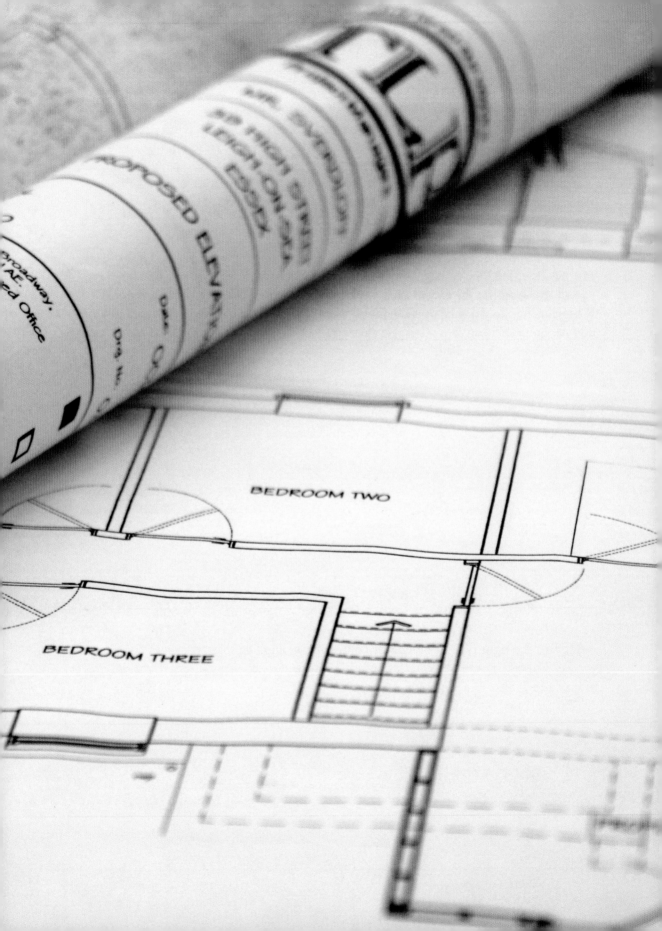

Chapter 3

planning your project

Preparing packages for obtaining quotes from builders

When considering who you will be inviting to tender for your project, it is important to remember that in order for you to compare competitive quotes, the builders must each be given a copy of the same drawings, specification, engineer's details and any other supporting material.

Your research of potential builders needs to be comprehensive in order to avoid your time being wasted. You need to satisfy yourself that you are sending the details of your project to the right type of builder or contractor. Compiling these packages takes time, money and effort, and this can be wasted if you send them to the wrong builders in the first place.

However, it is advisable to send the packages out to a sufficient number, as it is unlikely that all of those whom you ask to quote will actually respond. It is also advisable that you place a time frame on when you expect the estimate or quote to be returned. And make a list of who you will be inviting to quote, and take contact details so that you can chase them up if need be.

Most builders will want to visit the property to assess any logistical problems that may exist, such as where they will be able to place a skip and where the material can be unloaded to before it is used. The more information that the builder requests from you the more he is indicating his real interest in the project.

> **You should expect** the builder to visit the property to see where it is located, and therefore where the work will be carried out. Some projects (such as loft conversions, extensions and refurbishments) may require the contractor to carry out their own measurements and inspections – which may bring to light a few elements that have not been included in your specification. If this happens, you need to ensure that these form part of the written quote.

All will have a limit to the size of projects that they are able to carry out, which is determined by the labour resources and financial backing that they have available. Some will only carry out certain types of work that fit in with their area of expertise, for instance those who specialise in loft conversions and others who will only be interested in new build properties.

The nature of your project will dictate the type of builder you are likely to use, for example a small local builder working from home, or someone from a medium-size company with offices and staff. It is worth noting that larger companies with overheads may not be able to compete with the smaller builders' prices; however, their performance may be better. You can expect an established company to respond to you with a quote within one or two weeks, as they would normally have one person dealing with quotes.

Smaller builders may be able to respond in the same timeframe but if you do not make it clear when you would like your quote, they may take considerably longer than two weeks. They are usually running around to keep their projects moving and don't get as much time to spend on pricing projects as they would like. They may not even respond at all when they are under pressure on other projects.

Using specialist contractors

You now need to identify the elements of the project that you will be relying on contractors to carry out. This is where you should prepare individual packages and specifications for the different contractors to quote against. The more detail that you can provide in the package, the more easily the contractor will be able to identify the costs. You need to give them information on exactly what material you are supplying and what you will expect them to provide. You may need to take photocopies of parts of the drawings rather than provide a full set of drawings.

It can be tempting to request a verbal quote for a specific element of the work that appears to be basic, and may not warrant the time and effort to prepare a specification. However, verbal agreements are one of the biggest causes of disputes as they can be misinterpreted, and can lead to ill feeling and problems, particularly when the misinterpretation involves money. For this reason, even basic elements of the work need to be instructed or clarified in writing.

The main specification

In Chapter 1 we looked at the initial specification for giving the architect an idea of the thoughts a client may have on the whole project, and to clarify who would be responsible for particular elements such as supplying labour and/or materials. This information can now be used to prepare a much more detailed specification for obtaining detailed quotes or estimates.

The more detail you can put into your specification, the easier it will be for the builder or contractor to quote against. This will make it easier for you to analyse the costs.

A specification can be a simple document that supports a set of drawings and describes in more detail the scope of works and the materials to be used. On smaller projects the specification would be a supporting document, whereas on a larger project it would be the controlling document and would include four separate 'sections' detailing all other documentation. A detailed specification (controlling document) would start off with an index detailing each section, for example:

Section One: Preliminaries and General Conditions
Section Two: Specification/Scope of Works
Section Three: Summary and Form of Tender
Section Four: Appendices
1. A Architect's Drawings 01 – 22
2. B Engineer's Drawings 01 – 15
3. C Section Drawings
4. D Mechanical and Electrical Drawings

In order to construct a detailed specification, it is important to understand the reasons for specific inclusions and their significance. The following information will assist you in preparing your own specification, although this is a fairly comprehensive guide and you may not need to include all of the elements in the example. You could decide on using some of the elements in each section in order to produce a specification that suits your project. Each section will be broken down to explain some of the fundamental issues that need to be addressed. The specification will form part of the overall contract, and it will therefore be a useful document in the event of any discrepancies or misunderstandings. For the purpose of some of the following examples the specification is based on the demolition of a three-bedroom detached house, and the building of a four-bedroom detached house on the same plot.

The bullet point headings should be used; you need to specify in detail the exact type of material that you require, and where necessary the names and contact details of all suppliers and manufacturers. The drawings may give a general description of material but may not have specific details. It would be advantageous to refer to drawing numbers when detailing items within the specification. See Appendix (page 263) for an example of a completed specification excluding suppliers and manufacturers.

Section One: Preliminaries and general conditions

Section 1A – Project particulars

- Nature of work
  - Whether your project is a new build, extension or refurbishment it needs to be described as clarification of the project. For example:
    - Construction of detached four-bedroom house.
    - Refurbishment of semi-detached three-bedroom house.
    - Two-storey extension and loft conversion.
- Address
  - Although the address will be on all drawings, it is important that it is included in all contractual documents.
- Timescale
  - Specified number of weeks or best programme commencing from a preferred start date, which would be subject to confirmation.
- Client's name
  - All contractual documents need to include the client's full name.
- Name and address of Contract Administrator, if appropriate
  - Where a Contract Administrator has been appointed to act on behalf of a client in any capacity for all or part of the project, their full name and contact details need to be included.
- Name and address of Quantity Surveyor, if appropriate
  - Where a Quantity Surveyor has been appointed to act on behalf of a client in any capacity for all or part of the project, their full name and contact details need to be included.
- Name and address of Engineer, if appropriate
  - Where an Engineer has provided structural drawings for the project their full name and contact details need to be included.
- List of all Tender and Contract Documents, for example:
  - Specification: Sections 1–4
  - Site plans
  - Elevation drawings
  - Electrical drawings
  - Mechanical drawings
  - Drainage plans
  - Main drawings 1–22
  - Structural drawings 1–15
  - Sections A–A ~ B–B ~ C–C ~ D–D

Section 1B – The site and existing buildings

- Details of existing buildings on the site
  - Description of buildings (for example detached three-bedroom house with brick-build shed).
- Existing mains or services (normally part of appendices)
  - Drawings as supplied by statutory undertakers (for example, gas, electric or sewerage providers).
- Site investigation reports
  - If a site report has been prepared this would normally become part of the tender documents or attached in the appendices.
- Access to site via (specified route)
  - It may be appropriate to specify a preferred route particularly if there are special circumstances prevailing such as private roads or restrictions from other routes.
- Parking and any restrictions
  - If local parking restrictions are an issue, it may be appropriate to point out where contractors can and cannot park.
- Surrounding land and building uses
  - If the area is residential it is important to point out that minimum disturbance and consideration for neighbours needs to be a priority.
- Risks to health & safety
  - If there are any known H & S implications, for example contaminated land or hazardous material present such as asbestos, these need to be highlighted.
- Site visits
  - Site visits should be made by all parties tendering for the project to ascertain any potential problems/restriction that may have an effect on the execution of the work.

Section 1C – Description of the work

- The work
  - A more detailed description of the work to that in section 1A may be required, including other vital information such as the presence of existing services. For example:
    - Demolition of existing three-bedroom detached house with all existing foundations removed.
    - Building of a four-bedroom detached house on piled foundations.
- Completion by others
  - If there are elements of the work that are going to be carried out by the client or others, this needs to be highlighted, although this may also be referred to in other areas.
  - Examples of works being completed by others could include:
    - installation of kitchen

- installation of bathroom equipment/sanitary ware
- floor coverings.

⊖ Shrinkage and cracks
- The contract may detail defect liability periods and terms. However, you may wish to make it a condition of the contract within the scope of works that shrinkage cracks are to be made good after the defects liability period ends.

Section 1D – Contract of agreement

⊖ Type of contract
- There are contracts available off the shelf, which are designed to meet the needs of homeowners who are employing the services of builders or contractors. These are usually used on the larger projects.
- A formal contract needs to be drawn up for any project no matter how small (as detailed later in this chapter).

⊖ Earliest date of possession
- It is advisable to give a date of possession for the start of the contract, although it should be emphasised that it will be subject to confirmation.

⊖ Date of completion
- It is advisable to give an estimated time/amount of weeks for completion after the start date, but subject to acceptance, you could leave it to the best programme date that you receive with the tenders.

⊖ Extension of time
- You could specify here the only reasons that you are prepared to negotiate any application for extension of time. For example:
  - additional work by instruction
  - adverse weather conditions
  - design changes resulting in delays
  - delays caused by unforeseen circumstances but subject to discussion.

⊖ Penalty for late completion
- Penalty clauses are sometimes put in place on projects that are being undertaken with a formal contract. They spell out the legal details and are overseen by a Contract Administrator.
- Penalty clauses can become very contentious, particularly where changes and additional work have been carried out on a regular basis.
- Only impose a penalty clause if you fully understand the legal formalities of implementing one.

⊖ Defects liability period
- The period of defects liability needs to be set from date of Practical Completion; this could be 6–12 months from the date of Practical Completion.

⊖ Valuations
- A lender may dictate when valuations are carried out. As detailed in

Chapter 1, these could be linked to monies being drawn down at specific stages of the project.

- Depending on the duration of the project, you need to state the period between valuations; this could be, for instance, four weeks after the start of the project and every four weeks thereafter.
- This period may be reduced to reflect the nature and duration of smaller projects.

● Payment dates
  - These would normally be set at 7–14 days after the valuation.

● Insurance details
  - Copies of insurance certificates should be requested with the tenders in order to ensure potential builders or contractors have the appropriate cover to undertake your project.
  - Depending on the nature of your project there may be additional cover required other than that normally held by builders and contractors.
  - Your own insurance company should furnish you with the appropriate list of requirements.

● Disputes
  - In the event of a dispute arising that becomes difficult to resolve it is wise to discuss the appointment of an independent arbitrator prior to work starting on-site.
  - An independent arbitrator would normally be a member of the Royal Institute of Chartered Surveyors.

Section 1E – Tendering

● Scope of works
  - Only works as stated in the scope of works should be tendered for, including all preliminaries.

● Exclusions
  - Some builders or contractors may not have the resources for pricing or carrying out some elements of the work. If this is the case it is important that they let the client or Contract Administrator know as soon as possible with their reasons for the exclusion.
  - Some very specialist work may have to be organised by the client, in which case it is important to make it known that continuity and co-operation by all is expected.

● Acceptance of tender
  - You need to point out that no costs will be accepted by the client for the preparation of the tender, and no guarantees will be offered to accept the lowest tender.

● Period of validity
  - The period of validity of the tender should be stated; this would usually be between 12 and 16 weeks.

- Pricing of specification and clarification
  - It is important to invite those tendering to seek clarification on any issues that they may not fully understand.
  - All works in the scope of works and shown on the drawings should be sufficiently detailed to cover all works other than the exclusions noted above.
- The priced specification/scope of works
  - All elements that are individually described should be individually priced, otherwise it should be noted that they will be deemed as being included elsewhere.
- Errors in the priced specification
  - If it is found that there are errors in the drawings or specification, the builder must inform the client as soon as it becomes apparent.
- Programme of works
  - A proposed programme of works for the principal parts of the work should be requested from each party tendering for the project.
  - A tender may come in lower than the others; however, if the timescales are much longer it may be a disadvantage to you.
- Substituted products or materials
  - If the contractor intends to use alternative products or materials than those specified, the details must be submitted for acceptance by the client.
- Quality control procedures
  - It is important to know how the contractor will control the quality of all works and materials.
- Health & safety information
  - The information provided here will identify the contractor's commitment to providing a safe working environment.
  - The contractor's health & safety policy statement should identify the procedures and measures that are required to comply with The Health & Safety at Work Act 1974.

Section 1F – Documents, definitions and interpretations

- General qualification of wording
  - There will be specific terms used throughout the specification that need clarification in order to avoid misunderstandings.
- Additional documentation
  - Any additional costs to the contractor for the provision of additional drawings or other documentation must be noted.
  - Two copies of each working document would normally be provided.
- Dimensions
  - Scaling of dimensions from drawings is not normally acceptable. If any clarification is required the contractor should be directed to the drawing provider, or person in authority for making decisions of that nature.

- Documents and drawings provided by contractor/sub-contractors
  - Where documents and drawings are provided by contractors or subcontractors they may need to be submitted for approval to certain bodies depending on the nature of the work.
- Technical literature
  - All technical literature should be submitted and made available to the relevant parties at all times.
- Maintenance instructions and guarantees
  - All appropriate documentation of this nature should be specified, listed and made available to the client as and when requested to do so.

Section 1G – General management of the works

- Supervision
  - It is important to know what arrangements for supervision of the work will be put in place and that supervisory staff are experienced and/or qualified to sign off particular elements of the work.
- Weather conditions (keeping records)
  - There are particular elements of the work that cannot be carried out in extreme weather conditions, such as frost or heavy rain.
  - Delays can be caused due to severe weather conditions.
  - In order for the project to continue during adverse weather conditions, it would be expected that within reason the contractor has sufficient resources for providing suitable methods to minimise delays.
- Ownership of material
  - Any stripped out materials unless otherwise stated become the property of the contractor.
- Programme and progress
  - Arrangements must be made to monitor the progress of the work in relation to the programme.
  - A progress report should always be presented to the client.
- Meetings
  - A pre-contract meeting should always take place and timescales for subsequent meetings should be set between the main contractor and client.
  - The contractor should hold meetings with sub-contractors to discuss progress and any issues arising.
- Notice of completion
  - Where the project or parts of the project are coming to completion, the client should be given notice to carry out any independent inspections.
- Extension of time
  - Any claim for extension of time must be made well in advance of any completion dates previously stated.
- Cash-flow forecast

- An estimated forecast could be requested based on the programmed dates for valuations and the programme of works.
- Knowing approximately how much you will need to pay and when will help you to plan your finances.

Section 1H – Instructions and Information Flow

- Proposed instructions from client
  - A system for issuing instructions must be put in place in order to control costs (see page 178).
- Information required
  - The contractor should provide written requests for information where possible. All other verbal requests for information and subsequent answers of significance should be documented (see page 181).
- Measuring of work completed
  - Any work that is subject to a re-measure should be brought to the attention of the client.
- 'Day-work' agreements
  - Some work may be required which has not been priced. In order to keep continuity, a 'day-work' agreement may be a suitable arrangement.
  - Strict controls need to be observed for day-work operations.
- Interim valuations
  - A detailed breakdown of work must be provided to support a valuation.
- Unfixed material on-site
  - All unfixed materials on-site must be listed if included in the valuation.
- Listed off-site material or goods
  - There may well be expensive materials off-site that form part of a valuation, and for this verification should be requested.

Section 1J – Quality control and standards

- Good practice
  - All builders and contractors are expected to work within the scope of good building practice which should include the following:
    - consideration for others
    - being clean and tidy
    - working in a safe manner
    - being environmentally friendly
    - being responsible
    - being accountable.
- Materials – compliance
  - Unless otherwise stated, all materials should be new and meet British or European standards.
- Consistency

- Where materials are being delivered over a period of time it must be stressed that consistency in quality, colour, dimensions, density or any other specifics are checked to ensure the appropriate match.
- Protection of material
  - All material whether fixed, unfixed, boxed, wrapped, etc., should be fully protected during the project until Practical Completion has taken place.
- Workmanship
  - Since there are many types of trades and different materials it is expected that only experienced operatives will be used for specific elements of the work.
  - On particularly high-specification work, it should be emphasised that a high level of client monitoring may be exercised.
- Samples of material for approval
  - Samples should be requested for approval where the contractor is providing material that has been specified, but has not been physically seen.
- Samples of finished work
  - It may be necessary to request a sample of work to be carried out for approval to ensure that the quality of workmanship is equivalent to the quality of material and standards required.
- Setting out – discrepancies
  - If in the course of setting out or checking levels the contractor finds discrepancies, it should be emphasised that the client is contacted before any work proceeds.
  - Builders and contractors do come across these scenarios on a regular basis; however, the architect or client needs to make or at least be involved in all decision making.
- Appearance and fit
  - If in the event of a particular element being significantly different to that shown on a drawing or specification, the client should be informed immediately before work proceeds.
- Tolerances
  - Always remember that in construction some of the work may require a little more common sense applied when accepting part-finished work such as concrete floor levels.
  - There are standards on British tolerances for all manner of work; however, your own standards or those of independent inspectors should prevail.
- Services in general
  - All work affecting utility services must be in accordance with by-laws, regulations and statutory authority requirements.
- Mechanical and Electrical (M + E) tests
  - Apart from tests during construction phase, all M + E works should undergo a final test on Practical Completion.

Section 1K – Inspection and co-ordination of work

- Supervision
  - In addition to the general supervision of the main contractor's operatives, provision should be made to ensure that appropriate supervision is provided for all sub-contractors and their operatives.
- Co-ordination of building and engineering services
  - Sufficiently experienced personnel should be on hand to ensure the interfacing of trades.
- Access for inspection by client or Contract Administrator (CA)
  - Access must always be maintained for the client or CA to carry out inspections.
  - Notice must be given before removing access facilities such as scaffolding.
- Timing of tests and inspections
  - Appropriate notice should be given where specific tests need to be witnessed by the client, CA or engineer.
- Test certificates
  - Copies of certificates for any tests carried out should be submitted to the client or CA, and where appropriate a copy kept on-site.
- Proposal for rectifying defects
  - Where defects have been found, proposals for rectifying should be forwarded to the client or CA for approval.
- Quality control and test records
  - A system to ensure that all works are carried out in accordance with the specification should be put into operation.
- On completion of work – making good
  - By its nature, building work involves heavy equipment and material that can cause unplanned damage to property or areas outside the scope of works.
  - All making good of damage caused to the client's property or that of other property owners should be carried out at the contractor's expense.

Section 1L – Security ~ Safety ~ Protection

- HSE Approved Codes Of Practice
  - Contractors to comply with H & S Law.
- Security measures
  - Sufficient methods of security other than any specific requirements must be allowed for.
- Stability
  - The stability of the building and/or adjoining buildings must be maintained by the contractor at all times.
- Occupied premises
  - Additional care and health & safety measures will need to be taken in occupied premises.

- Proposals for safeguarding others should be submitted to the client or CA.
- ◑ Noise
  - There are regulations regarding noise. Any work that is likely to produce excessive noise needs to be adequately controlled.
- ◑ Pollution
  - Pollution can be caused by many types of work; it is important to point out any specific precautions that need to be taken.
- ◑ General nuisance
  - Nuisance can be seen as noise, smoke, dust, vermin, light fumes and many other sources; it is therefore important to highlight these, and in particular those that could cause a nuisance even in small quantities.
- ◑ Asbestos-based materials
  - If asbestos is known to be on-site it should be highlighted and dealt with in the appropriate manner.
  - If found during the project, it should be reported immediately to the client or CA.
- ◑ Fire prevention
  - Fire prevention on construction sites is very important. There are specific measures that need to be taken to ensure that the code of practice for fire prevention is achieved.
  - A fire plan and appropriate equipment should always be available on-site.
  - As part of the tender, a proposal for fire prevention should be submitted.
- ◑ Burning on-site
  - It is wise to exclude the burning of material on-site.
- ◑ Moisture
  - Moisture content in material can cause problems in many ways. Control measures need to be put in place to avoid moisture building up in materials, or for the controlled drying out process.
- ◑ Infected timber
  - This work to be carried out in the correct manner by the appropriate specialist contractor where certification is required.
- ◑ Waste
  - The control of waste needs to be planned and controlled to a high level.
- ◑ Existing services above or below ground
  - Special care needs to be taken to avoid damage to or disturbance of all existing services and where required specific drawings may need to be applied for.
- ◑ Roads and footpaths
  - The contractor must be made aware that the cost of damage to roads and/or footpaths will be their responsibility.
- ◑ Existing topsoil/subsoil/trees/shrubs
  - Specific care should be taken by all project contributors with regard to surrounding areas.

- Adjoining property
  - Unless otherwise agreed arrangements for access to adjoining properties for erecting scaffold or carrying out essential work should be made via the client or CA.
- Existing structures
  - The effects on existing structures should always be considered before any work is carried out.

Section 1M – Facilities and services

- Locations – site facilities plan
  - A plan should be requested to indicate the proposed site facilities and service runs.
- Meetings
  - Allowance should be made by the contractor for holding site meetings.
- Welfare/sanitary accommodation
  - Clean and maintained sanitary accommodation should always be made available for client and site operatives.
- Lighting
  - In order for the client or CA to inspect finished work, appropriate lighting should be available.
- Water – restrictions
  - Water is usually made available via the mains supply.
  - In the event of restrictions, alternative sources may be required.
- Telephone
  - A telephone link would normally be made available by the contractor to the person directly in charge of the site.
- Fax
  - In order for the flow of information to be sufficient to maintain continuity, a fax machine would normally be made available.
- Email
  - Email facilities would be desirable, subject to the contractor's resources and preferred methods of communication.
- Meter readings
  - All services would normally be charged to the contractor for units used in association with the project.
  - Arrangements for meter readings need to be made at both commencement and completion of the project.

Section two: Specification/scope of works

Section 2A – General information

- This section may include specific information that, although touched on in previous sections, may need more clarification. For example:
  - A fuller description of the overall work involved.
  - Clarification that health & safety information (such as the company H & S policy, together with Risk Assessments and Method Statements) will be required for specific elements of work.
  - The contractor will be responsible for costs incurred due to damage caused by delivery lorries or other associated parties.
  - Scaffolding work needs to have a separate programme for erection, duration and dismantling; all measures to protect the neighbouring properties from falling debris must be taken.
  - Details of what the programme should include.
  - The priced tender/specification should show each item individually priced as it appears on the summary page.
  - The highest standard of work will be expected, and where the client is not satisfied the contractor will reproduce the work at their own expense.

Section 2B – External works/demolition

- This section would identify the extent of work related to site clearance and demolition, and would list specific actions, illustrated in Appendix.

Section 2C – Excavations and foundations

- This section would clearly identify the extent of groundwork required and would include any specific actions to be taken as shown in Appendix.

Section 2D – Brickwork, block work and all associated work

- As the technical aspects of the main construction work are very comprehensive, your architect or other experienced person may be able to assist in the preparation of this section. However, if you are preparing the specification yourself, there should be sufficient information on the drawings for you to detail your requirements, see Appendix.

Section 2E – Suspended beam and block ground floor and associated work

- For details of the beam and block floor and associated work, such as insulation and screed, see Appendix.

Section 2F – Scaffolding

- Specific instructions regarding scaffolding need to be emphasised, in order to address any concerns you may have, see Appendix.

Section 2G – Roof structure, covering, fascia, soffits and bargeboards

- For details of the roof structure and finishing materials, see Appendix.
- If your roof has dormer windows, valleys or roof lights incorporated, these will need to be referred to.

Section 2H – Rainwater goods

- Type of gutters and down pipes should be specified with dimensions and colour, see Appendix.

Section 2J – External windows and doors

- External windows and doors would be supplied by a specific supplier (this is one area where the client could make considerable savings by providing them for the contractor to fix) or arrangements made for a specialist company to supply and fix.
- A door and window schedule would normally accompany the order and would be produced in conjunction with the drawings. For example:
  - door number G1 = front door; G2 = back door
  - window number G01 ~ G02 = ground floor windows
  - window number F01 ~ F02 = first floor windows.
- If you wish the contractor to supply and install the windows, you need to specify all technical details, see Appendix.

On 1st April 2002, all replacement glazing in dwellings came within the scope of the building regulations. Anyone who installs replacement windows or doors has to comply with improved thermal performance standards. The main reason for this change to the building regulations was the need to reduce energy loss. When a refurbishment or new development involves the replacement or installation of windows and doors, homeowners must ensure that they get a certificate from the local authority Building Control or have the work completed by a FENSA (Fenestration Self-Assessment Scheme) Registered Company.

FENSA was set up by the Glass and Glazing Federation (GGF), with Government approval, in response to the new building regulations for England and Wales.

Section 2K – External drainage

- Although the drawing will indicate where the drain runs are, a fuller description may be required that specifies the type of material to be used, see Appendix.

Section 2L – External paving and driveway

- ◒ It is important to clarify the type of blocks and make-up of base material when specifying any paving or driveways.
- ◒ It would be worth obtaining makers' recommendations and section details in order to ensure that the work is carried out to the specification, see Appendix.

Section 2M – Internal walls and floors

- ◒ Sizes of floor joists and direction of span would be indicated on the drawing. Any doubling up or bolting would also be shown. However, there may be some details that require clarification. For example:
  - joists to be strapped to wall at 1.200mm centres
  - solid strutting
  - end block to last joist
  - internal walls can be constructed from timber or blocks. This would normally be indicated on the drawing; however, the width may need to be clarified where they are not supporting walls, see Appendix.

Section 2N – Internal doors and ironmongery

- ◒ Architect's drawings usually indicate the doors with a reference code. For example: DG-01 indicates the front door (Door 01 on Ground Floor). Internal doors would follow on in sequence DG-02 ~ DG-03, etc.

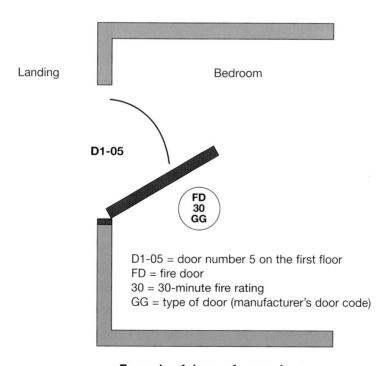

Landing

Bedroom

D1-05

FD 30 GG

D1-05 = door number 5 on the first floor
FD = fire door
30 = 30-minute fire rating
GG = type of door (manufacturer's door code)

Example of door reference key

- Doors on the first floor would start off in sequence with the first door code being DF-01 (Door 01 on First Floor), DF-02 ~ DF-03, etc.
- Some architects will produce a door schedule as a matter of course in order to detail the type of door and associated ironmongery, see Appendix.
- The door schedule would normally be issued as an appendix to the specification.

Section 2P – Kitchen units and fittings

- Builders can arrange for kitchen supply and installation; however, this is an area where you could make considerable savings as builders could mark this element up with their profit margin.
- The builder would be expected to provide all services to the kitchen including:
  - gas
  - electric
  - hot and cold water
  - waste pipes.
- If you are planning to have an extractor fan installed, the associated work would normally be carried out by the builder. For example:
  - providing the power and preparation of the hole for the ducting through the wall.

Section 2Q – Staircase and associated works

- Stairs and associated works would normally be carried out completely by the builder.
- There are many different types of newel posts, newel post caps, spindles and handrails. It would be necessary to source the appropriate type and specify them in detail.

Section 2R – Plasterboard, plastering and coving

- You will need to specify the thickness of render, plaster and plasterboard and how many layers are required.

Section 2S – Plumbing, heating and sanitary fittings

- Plumbing and heating specifications can be very comprehensive. They need to be designed by the appropriate professional mechanical engineer to achieve maximum efficiency, and to comply with relevant regulations.
- Plumbing contractors should be aware of the regulations they need to comply with; however, it is worth pointing out specific elements such as:
  - type of pipes to be used (copper or plastic)
  - type of radiators
  - type of towel rails.
  - isolation valves to be fitted and easily accessible to all hot and cold supplies that feed sinks, basins, sanitary fittings, appliances, to facilitate future maintenance.

- All pipe work to be clipped/fixed to manufacturer's recommendations.
- If you are going to leave the design and installation of the heating system to builders, you need to ensure that they are suitably qualified to do this.
- At this stage it is worth considering if you will be developing the property at a later date, and if so pointing out the need for the boiler to be sized accordingly.
- Where appropriate, copies of certificates will be required.

Section 2T – Electrical installation and fittings

As referred to in Chapter 1, Part P of the building regulations deals with Electrical Safety. This regulation came into force on 1st January 2005. It requires companies and tradesmen who undertake electrical installations to certificate their work and notify it to the local authority Building Control department. It has been introduced to reduce the number of deaths and injuries arising as a result of defective or unsafe electrical work. Most electrical work in houses including routine work (such as adding a new socket) must now, by law, be notified to the local authority Building Control department. This is not only confined to work carried out by qualified electricians, but also householders who carry out DIY jobs. General tradesmen who undertake electrical work in the course of their normal activities (such as bathroom and kitchen fitters) also need to have their work certified or inspected.

There are currently two methods to prove compliance:

- The householder or electrical contractor can notify the relevant local authority Building Control department, obtain approval, and pay the relevant charges for every installation. The charges may vary between areas, but at the time of going to press are in the region of £50 to £250 for every installation inspected and certified.
- The electrical contractor can register with a 'Competent Persons Scheme' such as ELECSA (Electrical Self Assessment) to self-certify that their work complies with Part P of the building regulations. The notification to the local authority is then handled by the organisation (e.g. ELECSA) running the scheme.

As with plumbing and heating, the electrical specification can be very comprehensive. It needs to be designed by the appropriate professional electrical engineer to achieve maximum efficiency, and to comply with relevant regulations.

All electrical work should be carried out in strict accordance with the current IEE (Institute of Electrical Engineers) Wiring Regulations.

Section 2U – Floors and finishes

In general, finishing work could be left out of the specification completely and organised by the client at a later date, once it is known how much of the contingency sum has been spent. The quality of finishing material may need to be lowered if the contingency sum has been exhausted or, alternatively, the quality could be raised if the project has gone smoothly.

- By arranging these elements independently from the main building works, it is possible to cut costs. The builder may use specialist contractors to carry out the work.
- This type of work is classed as specialist work, although depending on the quality of the material and standard required, it could be carried out by general tradesmen.
- It would be worth obtaining several independent quotes for floor finishes, as well as quotes from builders.

Section 2V – Ceramic wall tiling

As stated above for Floors and Finishes, the quality of material could be reviewed nearer to the end of the project when the main building costs have been ascertained.

Section 2W – Internal decorations

- It is advisable only to carry out painting work to walls and ceilings rather than have wallpaper applied, as the drying out and settlement of new work can take several months.
- This is, of course, an area in which considerable savings can be made by clients who have the ability to undertake the decorating work.
- If you are planning to carry out the decorating work, it is important that the joinery work is rubbed down, with knotting and primer applied prior to being fixed in position.

Section three: Summary and form of tender

The specification would normally be presented in a format in which the contractor can fill in the costs next to the item, and at the bottom of each page the total would be shown. The totals from each page could then be transferred on to a summary sheet. The following examples show how the specification could be presented, and how a summary page of the specification would look.

Contractors sometimes price for elements that may appear elsewhere in the specification. In this case they would normally write 'included elsewhere'. If an item is not priced and is not qualified, it is deemed to be included elsewhere.

SAMPLE SPECIFICATION

Section 2B – External works/Demolition

1.	Prior to demolition work make arrangements for an asbestos identification survey to be carried out, to include a report for safe removal to a licensed disposal site. Allow for removal and carting away (provisional sum)	
2.	Prior to demolition, liaise with the appropriate statutory undertakers to locate and isolate all services, i.e. a. Gas, b. Electricity, c. Water, d. Communication networks	
3.	Submit to client and local authority Building Control a method statement for the safe demolition of existing buildings	
4.	Protect trees, surrounding fences and brick walls identified on drawing No …	
5.	Erect temporary hoarding to secure the site during the demolition stage	
6.	Carefully remove and retain for re-use all ornamental fireplaces	
7.	After safe removal of asbestos-based material carefully demolish and cart away existing building including: a. Roof covering and structure, b. Windows, doors and frames, c. Floor joists, d. All brickwork, e. Foundations and slab	
8.	As indicated on drawing, remove trees, shrubbery and foliage, grub up roots and cart away	

Section 2C – Excavations and Foundations

1.	Excavate to reduced level as shown on drawing and put to one side any clean topsoil for re-use by client; all other debris to be removed from site	
2.	Allow for protecting all services such as: a. Drain runs, b. Gas mains, c. Water mains, d. Electrical services	
3.	Carry out piling as per engineer's drawing to depths shown and report to client if depths are exceeded with the additional costs on a pro rata basis	
4.	Construct ground beams in accordance with engineer's drawings	

Section 2D – Brickwork, Block work and all Associated Work

1.	As indicated on drawing No … construct cavity wall from ground beam up to damp-proof course level using class B engineering bricks	
2.	Damp-proof course to be bitumen-based Class A and lapped 110mm where necessary with adhesive as per manufacturer's recommendations	
3.	Provide sample of face bricks as specified including a built sample of 1.0 x 1.0m for approval prior to main construction work	
4.	Construct main walls with a weathered struck mortar joint and 75mm cavity filled insulation as identified on drawing No …. Inner leaf of cavity wall to be constructed of 100mm standard blocks as identified on drawing No …	
5.	Allow for expansion joints where indicated on drawing No …	
6.	Build in stainless steel vertical twist type wall ties at appropriate dimensions i.e.: A. 900mm centres horizontally and 450mm vertically B. 225mm centres vertically at all openings and within 150mm of opening	
7.	Allow for all openings as indicated on drawings with lintels as specified on lintel schedule, minimum 150mm bearing	
8.	Cavity trays and weep holes in accordance with manufacturer's recommendations	
9.	Where indicated on drawing fix telescopic air vents at 1.800mm centres	
10.	Cavity closers type … to be installed as per drawings at all openings	
11.	Spray weed killer to the entire internal area of the new building	
12.	Construct 200mm sleeper walls in accordance with engineer's drawings, built off from main ground beams	
PAGE …	TOTAL

ITEM	SECTION	COST
	SUMMARY OF TENDER	
1	1A - 1M PRELIMINARIES AND GENERAL CONDITIONS	
2	2B - EXTERNAL WORKS/DEMOLITION	
3	2C - EXCAVATIONS AND FOUNDATIONS	
4	2D - BRICK AND BLOCK WORK	
5	2E - SUSPENDED BLOCK AND BEAM FLOOR	
6	2F - SCAFFOLDING	
7	2G - ROOF STRUCTURE, COVERING, FASCIAS, etc.	
8	2H - RAINWATER GOODS	
9	2J - EXTERNAL WINDOWS AND DOORS	
10	2K - EXTERNAL DRAINAGE	
11	2L - EXTERNAL PAVING AND DRIVEWAY	
12	2M - INTERNAL WALLS, FLOORS, DOOR FRAMES	
13	2N - INTERNAL DOORS AND IRONMONGERY	
14	2P - KITCHEN UNITS AND FITTINGS	
15	2Q - STAIRCASE AND ASSOCIATED WORK	
16	2R - PLASTERBOARD, PLASTERING AND COVING	
17	2S - PLUMBING, HEATING AND SANITARY FITTINGS	
18	2T - ELECTRICAL INSTALLATION AND FITTINGS	
19	2U - FLOORS AND FINISHES	
20	2V - CERAMIC WALL TILING AND MIRRORS	
21	2W - INTERNAL DECORATIONS	
	TOTAL	

Form of Tender

The Form of Tender is a single form/document that is used to accompany the tender documents. The form is used specifically for the contractor to enter the final price. It would include information such as how long they estimate the project will take, how much time would be needed to mobilise on receipt of order, and that they confirm the price is valid for at least three months from date of tender. This information would be used in conjunction with the terms of contract. Opposite is a sample Form of Tender document that shows the type of information that may be included.

FORM OF TENDER

Client Details

NAME:	
ADDRESS:	
DATE:	
PROJECT:	
TENDER TO REACH THE CLIENT BY --/--/--	

We the undersigned hereby offer to undertake and execute to completion the works as specified in the drawings, specification and conditions of contract. We understand that all work will be accepted for payment on satisfactory inspection by the client or Contract Administrator, for the Firm Price Tender sum of:

(£   ) + VAT where applicable

We will require (   ) weeks to complete the work on-site as per our programme and will be available to commence the planning of works on receipt of your order and written instructions

We will be able to commence work on-site (   ) weeks after receiving instructions

The period of validity of this tender will be three months from the date above

being --/--/--

This Tender is intended to be competitive and has not been fixed or adjusted in accordance with agreement or arrangement with any other person

The building owner/client reserves the right to accept or refuse tenders from those submitted

For and on behalf of: (Builder/Contractors details)

NAME:	
POSITION:	
ADDRESS:	
TEL. NO:	
DATE:	
SIGNATURE:	

Appendices

The appendices would be listed in the tender documents list and would typically include:

A. Architect's Drawings
B. Engineer's Drawings and Specification
C. Section Drawings
D. Mechanical and Electrical Drawings
E. Product Ref No:... Technical information and data sheet

Where specific manufacturers have been included, it may be appropriate to include details or technical information for the product in question.

Quotes/Estimates

The specification for large projects needs to be comprehensively presented for accurate tenders to be obtained. However, smaller projects do not lend themselves to the same level of preparation. If you have prepared a basic specification to back up drawings for a small project, you may well be in a position where you will receive a price for a project that is not broken down to show the different elements. In this instance, you would need to sit down with the contractor to discuss the scope of work in detail to ensure that all elements have been priced. Any verbal discussions and subsequent agreements should be recorded on paper with signed copies for each party.

When you start to receive quotes or estimates, bear in mind that an *estimate* for work is different from a quote. In order to meet the terms of any contract, it is important to understand why some prices for work are termed as estimates and others as quotes.

Estimate

An estimate is seen as a statement of the expected/approximate cost of a project or task, which should specify the elements that have been allowed for in terms of labour and/or material. Because an estimate is only an approximate cost, it is likely that costs will rise. Estimates are usually provided where there may be elements of work that are difficult to ascertain fully in regard to the level of labour or material content required. Where a detailed survey has not been carried out due to difficulties in obtaining access, or a more detailed investigation is required, it is not always appropriate to provide a fixed quote.

Examples of this are as follows:

- a refurbishment project that involves carrying out work such as hacking off and replastering sections of walls or ceilings that have been roughly specified (other than the complete wall or ceiling)
- external ground works where the ground condition is not fully known; if the estimate was based on excavating 10 cubic metres of earth, and it transpired that the ground consisted of old footings and concrete that had to be dug out by hand, this would clearly lead to a substantial amount of additional time and resources
- stripping off several layers of wallpaper back to the plaster and making good, ready for redecorating
- replacing defective floor joists where dry rot or woodworm are present.

As you can see from these common examples, it is important to discuss the details of any estimate and to have an idea of the additional costs that may be involved. Agreement for additional work could be based on hourly, daily, or square metre rates, with materials invoiced separately. Costs for specialist equipment that may be required also need to be identified. There are many terms of contract that can be applied, but it is important to establish these before the work starts.

Your specification needs to be as comprehensive as possible in order for you or a contractor to be able to understand the elements that need quoting for, and those that need to be qualified with terms such as 'estimate subject to investigation'. Where the extent of any element of the work is not fully known, you should consider allowing a contingency of about 10 per cent to cover any shortfall. If you suspect that you will find more problems, then a much higher contingency may be required for certain elements.

If you agree to have work carried out based on an estimate, all additional work needs to be sanctioned before being carried out. All work needs to be fully documented in order for you to know where the additional costs have been spent. It is therefore a top priority to ensure that you (or your representative) have the authority to issue instructions for additional work, and to monitor the work. If you are presented with an invoice for additional work, it is essential that you have evidence of the work having been carried out, particularly if it is work that is going to be covered up or removed from site. This is where photographic evidence is useful, and you could request evidence of additional work to be provided as supporting material for an invoice.

Quote (or quotation)

A quote or quotation is seen as exact information on the price of a commodity or service. In the case of a quote for a building project or part of a project, the information would be based on the information that was provided by the client. In order for a builder or contractor to provide and stick to a quote, the information would need to have been fairly comprehensive. Good specifications and drawings are fundamental for builders and contractors to provide quotes to clients, as opposed to estimates. Where drawings and/or specifications are not clear or do not have sufficient information, a builder or contractor may ask for clarification on certain points or they may contact the architect or engineer. Only if they receive clear answers will they normally be prepared to provide a quote. It would be difficult to hold a contractor to a quote if the wording for specific elements included terms such as 'subject to a re-measure'.

Where drawings and/or specifications are clearly inadequate and have many questions that need answering, the contractor may be deterred from providing a quote. In this case, the builder or contractor may only be prepared to provide an estimate, which would be subject to re-measures or confirmation of costs upon further investigation, or precise surveys. In many cases where drawings and/or specifications are poor, builders may provide a high quote having calculated the worst-case scenario. It is not uncommon for builders to win projects having priced way over the odds for the project, because other builders who were invited to quote would more than likely have done the same.

If you find that the quotes or estimates are coming back far in excess of your initial budgets, you may need to consider employing the services of an independent surveyor. On larger projects this is a good option, as the potential for paying over the odds for some elements of the work are greater. A professional quantity surveyor will not only have the experience to know what you should be paying for each element, but will also be able to oversee the payments and contract terms.

Whatever project you are undertaking, it is possible to identify elements that you will be able to obtain quotes for, and those that will be estimated. There are few major uncertain costs that are associated with a new build project. However, works below ground can stretch the budget considerably. This is due to the fact that while ground conditions are generally known, it can never be taken for granted that *assumed* ground conditions are correct. Additional costs to a new build project can also be incurred due to severe or extreme weather conditions. This can cause problems and delays during the building process.

Regarding extensions, loft conversions and particularly refurbishment, the unknown factors are greater in number, although on the whole they are less likely to be individually problematic with regard to cost. It is the amount of combined additional work that usually causes budgets to be overstretched.

> **It is very important** when working out your budget that sufficient contingencies have been allowed; this may not only help to keep the stress levels down but it may also give you a pleasant surprise if all goes well.

'Like-for-like' quotes

As mentioned in Chapter 1, if you are going to use one builder/contractor for the whole project, you need to be sure when you send out your specification, drawings and details for quotes, that they are capable of carrying out the work. Your list of builders/contractors needs to include five or more, as some builders will show interest in quoting for your project, but when they see the drawings or location they may decide that they do not want to do the work. They may not let you know, and while you are waiting for a response, time is ticking away.

By sending out four or five sets of identical information, you stand a chance of getting at least a couple of detailed quotes back. You need to be firm when issuing information for quotes and put a timeframe in place for when you would like the response. You may have to follow up your request for a quote with phone calls. If you find that you are not getting a satisfactory response from a particular builder who initially showed interest, this could be a warning sign that the level of service when carrying out work may be the same.

In order to receive 'like-for-like' quotes, it is important that you provide the right information and paperwork for the builders or contractors to demonstrate how the price has been put together, as per the specification and summary sheet.

Breakdown of quote

The quality of information that you are likely to get back from builders or contractors will depend on the quality of information that you are providing. If you do not have the resources to put together a comprehensive specification or if your project is fairly small and doesn't warrant the effort, you will at least need to provide a formatted sheet, which breaks down the elements of the project. This sheet could simply identify all of the elements that need to be priced. It could indicate the elements that you will be responsible for providing or undertaking yourself. If you are planning to use one builder for the whole project, the sheet will enable you to see how each builder has built up the price. The quotes will, of course, vary but it is not wise to accept the cheapest one before analysing others.

Once you start breaking down the quotes, it may highlight the areas in which a particular builder's quote has come in considerably lower. If the difference in cost is down to a lower standard or different type of material to that which has been specified, this will need to be discussed and either justified, or costs adjusted to suit.

If a quote that you have been presented with is considerably cheaper than others, it may well be that the company or person has much lower overheads or has a lower mark-up on the profit margin. It would not be unreasonable for you to make detailed enquiries into how the price has been put together. The most important factor is that you must be satisfied that the particular project or task that has been quoted for can and will be completed satisfactorily within the price. If you do not feel confident in the ability of a company or person, or if you feel that their explanation is a bit 'loose', it would be advisable to consider carefully whether to enter into a contract with them.

If a builder or contractor quotes for and wins a project and subsequently realises that mistakes have been made in their calculations, they may try to find ways to increase the costs. In this instance it has the potential to cause problems between them and the client as the client would have accepted the quote in good faith. Even when a legal contract is in place, builders and contractors will try to find ways to keep the costs down if they have underpriced a job. This may lead to them cutting corners and using inferior labour or materials.

In extreme cases, they may pull out of the project completely if they stand to lose too much money by sticking to their commitment. Needless to say, if they are owed a considerable amount, the likelihood of them pulling out would be less. If a builder or contractor did pull out of the project, the incoming builder would probably charge more than the normal rate to take over. When a situation like this arises it creates an air of suspicion, and all of the work that had been previously carried out by the outgoing builder would have to be fully checked. In order for a new builder or contractor to sign off electrical or mechanical work that may have been carried out by another contractor, they could insist on completely removing and replacing all previous work.

If a genuine mistake has been made it may be beneficial to negotiate the way forward rather than forcing a builder to lose money. In this instance, you would benefit from employing a professional surveyor to negotiate on your behalf.

As you will see in the sample summary sheet for a small project (overleaf), this will allow the contractor to break the elements down in order for you to compare costs. You may meet some resistance when requesting a breakdown quote, as some of the elements may be measured as a square metre rate for supply and fix. However, you will still be able to

SAMPLE SUMMARY SHEET				
Description	Material	Labour	VAT	Total
1	Foundation to DPC				
2	Brick/block to roof				
3	Roof and windows				
4	Mechanical & electrical				
5	Plastering				
6	Carpentry & fitted elements				
7	Bathroom	Client			
8	Kitchen	Client			
9	Decorating & wall tiling				
10	Floor finishes	Client	Client		
11	External works				
12	Contingency				
	Totals				

compare prices by looking at the overall cost. It is in your interest to know the true cost, as you may find that the quote has come in under your budget, which would give you the opportunity to either upgrade the specification or have other works carried out.

Being able to analyse specific costs where quotes have come in higher than expected will enable you to decide on areas where you may be able to make amendments to the specification. You may be able to remove from the contract some of the work such as decorating, which you can carry out yourself or leave until you have arranged sufficient funds.

Many builders will be prepared to negotiate on a quote if it means that they will get all of the work. Builders may at times be prepared to carry out some or all of the work at cost in order to keep their workforce employed, particularly if they have a slack period. In this situation, the quote would come in considerably lower than others, and the builder may be happy to inform you that the costs are low because the order books are down.

Assessing quotes

If you have provided the same information to several builders, it does not necessarily mean that you will receive the same information back from each one. Filling in blank spaces on a specification with figures and returning them to you as a completed quote is not all you should be looking for. Depending on the quality of the drawings and specification that you have provided, you should expect to receive some questions or clarification of particular points. It is unusual for an architect and a client to provide every detail for a builder to quote a project accurately.

Some builders will assume an answer to a question in the event of them being unsure about small discrepancies. Builders that ask for clarification are probably pricing the project correctly rather than guessing at some elements. If a builder is particularly interested in winning the project, you may receive details of some of the materials that they propose to use with the quote. You may also be provided with ideas for making cost savings by using different types of materials of the same quality as those specified, but which cost a good deal less.

There are many other factors that should be considered when assessing quotes, for example:

- Did the quote arrive on the due date?
- If the quote did not arrive on the due date, was there an explanation?
- Were all requested documents attached to the quote?
- Was the quote presented in a professional manner?
- Was there a covering letter explaining any specific details?
- Was there a general covering letter?
- If details were handwritten, were they legible?
- Did the builder visit the site to ascertain potential problems?

Deciding on which builders or contractors to use

One of the most important decisions you are likely to make when undertaking a project of any size is that of employing the services of a builder or contractor. These decisions can be narrowed down by making the right enquiries and only inviting appropriate builders or contractors to quote in the first place. Since the investment that you are making is probably substantial, it stands to reason that you must protect it as far as is reasonably practical. You may have made tentative enquiries initially to ensure that the builders you have invited to give quotes are experienced in the type of work that you are proposing. However, before making the all-important decision of awarding the contract to a specific builder, it would not be unreasonable to ask to see work on which they are currently engaged. Seeing at first hand how they treat the client's property and how they work will give you an idea of what to expect if you do decide to employ them.

Some tell-tale signs of good building practice and responsible behaviour would be that:

- vehicles associated with the site are parked sensibly
- the site on first appearance is clean, tidy and organised
- roads and pavements are not damaged or muddy
- where appropriate, suitable fencing is available
- the site facilities are appropriate for the project
- workers' appearance is in keeping with a professional outfit
- equipment is clean and in good order
- no trailing leads are evident, except in close proximity to the work
- standard of work appears to be good
- workers are polite
- noise levels are acceptable (no loud radios)

- security and health & safety signs are evident, i.e.
  - visitors to report to office
  - there are warning signs for pedestrians that construction work is in progress
  - standard health & safety signs.
- workers are wearing Personal Protective Equipment
- the company vehicles are well presented
- a company sign is on display
- where appropriate, material is neatly stacked and protected.

Some tell-tale signs of bad building practice and irresponsible behaviour would be:

- untidy site
- materials poorly stacked and unprotected
- lack of signs generally
- loud radios
- poor site facilities
- workers not wearing Personal Protective Equipment
- poor standards of work
- signs of burning material on-site
- inadequate security, e.g. no fencing (if appropriate).

Once you have had the opportunity to look at the quotes, it would be worth discussing them in detail with each of the providers. It is not only the paperwork and price that give you information for deciding on which builder or contractor to use, it is also important to look for signs that the builder is client orientated. If a builder appears to be run off his feet and does not have time to discuss the detail of the quote, this could be seen as a sign that there is a lack of organisation or real interest. Whether a builder is busy or not, it is important that a senior representative of the company who has knowledge of the project is able to demonstrate why certain elements are more expensive than in other quotes.

If a quote has come in considerably lower than others but the builder does not show real interest, or if you have taken the opportunity to look at other projects that they are running and don't like what you see, don't be tempted to take a chance on things working out. Once you enter into a contract to employ the services of a particular builder or contractor who under-achieves, it may be very difficult to get them to work to your required standard. This will obviously have a negative effect on the relationship, and has the potential to cost a great deal more in the long run.

It would be much more advisable to negotiate with a builder whose quote has come in higher, but who is client focused and can demonstrate that their work is of an acceptable standard.

Contract of agreement

Building work (or specialist contract work) normally involves sums of money that warrant some sort of control mechanisms to be put in place, in order to protect the investment, in the event of problems. A verbal agreement may seem to be all that is necessary for a small building project, such as a roofer or heating engineer would undertake. If you have used a particular builder or service provider before and have been satisfied with the arrangements and standard of work, you may be even more against the idea of implementing a contract of agreement. This is understandable – after all what can go wrong?

Many of us will have heard the stories and seen on TV the misery that people suffer when things do go wrong, even with small jobs. By preparing a contract or agreement yourself, or buying a contract 'off the shelf', it is possible to have the work done with peace of mind. A legally binding contract with sufficient detail will enable you to put your case in court in the event that things deteriorate to that level. Whatever size project you are preparing to undertake, it is of the utmost importance that a written agreement of some description is prepared and signed by the client and the contractor.

Let's look at off-the-shelf contracts and how you can prepare a simple agreement yourself. Whatever contract you decide to implement it would be worth having your legal representative check it over once it is prepared.

The Joint Contract Tribunal (JCT) provide off-the-shelf contracts for the construction industry. For domestic building work there are two main contracts:

- Building Contract for a Home Owner/Occupier who has not appointed a consultant to oversee the work.
- Building Contract for a Home Owner/Occupier who has appointed a consultant to oversee the work.

Let us look at Building Contract for a Home Owner/Occupier who has not appointed a consultant to oversee the work.

The contract is presented in a pack/folder and contains:

- a sample covering letter for sending out to potential contractors
- two building contracts:
  - one for the builder
  - one for the client
- guidance notes.

The documents are very easy to understand and they don't have technical or legal jargon that would require a solicitor to decipher.

The contracts themselves are identical in every way, except that one has 'Contractor' on the front and the other has 'Customer'.

The contracts are in two parts: Part 1 – The arrangements for the work, and Part 2 – The Conditions. Each part is divided into separate sections, covering the specifics of the contract.

Part 1 – The arrangements for the work cover:

A. The work to be done
B. Planning permission, building regulations and party walls
C. Using facilities on the premises
D. Price
E. Payment
F. The working period
G. Product guarantees
H. Insurances – before the work starts
I. Working hours
J. Occupation and security of the premises
K. Disputes

The following is a brief description of what each section covers:

A. The work to be done

- A short description of the work to be carried out.
- List of documents and dates, e.g.:
  - contractor's quote
  - drawings
  - specification
  - other documents.

It is advisable that when using this contract all documents that change hands should be initialled by the contractor and client. The client keeps the original documents.

B. Planning permission, building regulations and party walls

- This section has tick boxes that indicate who is responsible for obtaining the appropriate permission regarding planning permission, building regulations approval and party wall consents.
- There is a clear instruction here stating that the builder cannot start work before the appropriate planning permission and party wall consents have been received. However, it does state that the contractor *can* start work without building regulations approval provided the local authority are notified at least 48 hours before work commences.

C. Using facilities on the premises

- There are four tick boxes in this section indicating the facilities that the contractor can use free of charge, e.g.:
  - electricity
  - telephone/fax
  - washroom/toilet
  - water.

D. Price

- ◉ The price quoted for the work.
- ◉ States that the contractor will itemise the price and show where VAT is to be added and at what rate.
- ◉ Indicates whether the contractor is responsible for local authority costs.
- ◉ Specifies that the contractor is responsible for unexpected costs that could have been ascertained prior to pricing.
- ◉ Indicates that the price will be revised where changes are made to the specification, whether this is an increase or decrease in cost.
- ◉ The price after revisions due to changes will be the total price.

E. Payment

- ◉ This states that the client will pay either 95 per cent of the work on completion or by instalments at agreed stages of the contract up to 95 per cent of the contract.
- ◉ The client will pay the remaining 5 per cent upon satisfactory resolution of defects, up to three months after the work has been finished.

F. The working period

- ◉ There are two tick boxes here which state that the contractor will either:
  - ▪ start the work no later than —/—/— and complete it by —/—/—
  - ▪ start the work which will be complete within ( ) weeks from a date agreed between the client and the contractor.
- ◉ The work is complete when both parties agree that the main works (including changes) have been carried out.
- ◉ The working period can be extended in certain circumstances subject to other conditions being met.

G. Product guarantees

All manufacturer guarantees to be issued to the client for products installed in the work.

H. Insurances – before the work starts

- ◉ The client will inform the household insurers of work about to start.
- ◉ The contractor must have 'all risks' insurance.
- ◉ The contractor must have 'public liability' insurance.
- ◉ There is a box for the contractor to show the amount insured for any one claim.

I. Working hours

The client will show here the hours that the contractor will work from Monday to Friday, although it is not unusual to agree alternative hours once the project starts.

J. Occupation and security of the premises

- There are tick boxes here for the client to identify whether the premises will be occupied.
- Specifies that the contractor must take practical and common sense precautions to deter intruders from entering the property if the premises are unoccupied.

K. Disputes

- Either party can start proceedings in the event of a dispute.
- Indicates a method of resolving disputes by the appointment of an adjudicator.
- Clarification of adjudication process.
- Clarification that costs associated with adjudication will become part of the contract.

Part 2 – The Conditions cover:

1. Contractor's responsibilities
2. Customer's responsibilities
3. Health & safety
4. Changing the work details
5. Extending the working period
6. Payment
7. Contractor's continuing responsibility
8. Bringing the contract to an end
9. Insolvency
10. Other rights and remedies
11. Law of the contract

1. Contractor's responsibilities

- Carry out the work as stated competently.
- Use new materials of a good quality, unless otherwise agreed in writing by the client.
- Keep to the start and finish period as stated.
- Keep continuity.
- Inform the client of sub-contracted work.
- Keep tools and equipment stored away.
- Keep site clear of rubbish.
- Be responsible for any damage caused to the property or neighbour's property.
- Leave all working areas clean.
- Maintain legal responsibilities.

2. Customer's responsibilities

- Allow access for the contractor to carry out the work.
- Not to obstruct the working area.
- Allow the contractor to work to their sequence.

3. Health & safety

- ◉ The contractor must take all necessary steps to:
  - prevent or minimise risks to others
  - be environmentally aware
  - make provision for temporary weatherproofing.

- ◉ The client must:
  - take notice of the contractor's health & safety (H & S) instructions
  - ensure that unauthorised people do not enter site/work areas, particularly children.

4. Changing the work details

Only the client has the authority to change the work details.

- ◉ If the client increases the work shown, the contractor must have costs agreed before proceeding.
- ◉ If the client reduces the work shown, the contractor must adjust the price accordingly.
- ◉ If any changes alter the cost but not the work involved, appropriate adjustments must be made.

5. Extending the working period

- ◉ The client will agree to extending the work period if:
  - extra work has been added or changes have been made
  - circumstances beyond the contractor's control have occurred.
- ◉ The contractor can claim for extra costs where the client has delayed the work.

6. Payment

- ◉ The client will pay for work as completed, in accordance with item E under *The Arrangements For The Work*.
- ◉ The client will pay 95 per cent of the invoice within 14 days of receipt.
- ◉ The client will pay the remaining 5 per cent:
  - when the contractor has put right any defects
  - within three months after completion or at the end of the three months.

7. Contractor's continuing responsibility

For six years after completing the project, the contractor will be responsible for faults in the work caused by him, other than fair wear and tear.

8. Bringing the contract to an end

- ◉ If the contractor is not meeting his obligations, the client should give a written warning stating the matters concerned. If these matters are not corrected within seven days, the client can end the contract with a further written notice that will be immediately enforced. The main reasons for a client ending the contract would be:

- contractor not working on a regular basis
- health & safety or environmental responsibilities being ignored
- poor standards of work.

☻ In this situation any money owed will only be payable when the work has been completed by another contractor.

☻ If the client is not meeting their obligations, the contractor can send a written warning stating the matters of concern. If these matters are not resolved within seven days, the contractor can terminate the contract by issuing a further notice, which would come into force immediately. The main reasons for a contractor ending the contract would be if the client:
- does not pay, without good reason
- obstructs the contractor from carrying out work.

☻ In this situation the client will be expected to pay for work properly carried out and for materials that are on-site.

9. Insolvency

☻ If either party becomes insolvent (unable to meet debts) the contract will end, unless other arrangements are made by a professional insolvency practitioner to allow the contract to continue.

☻ If the contractor ends the contract due to insolvency, the client will only have to pay money owed to the contractor when the work is completed by others.

10. Other rights and remedies

☻ Claims can be made by either party for costs resulting from the terms of contract being broken by one or the other.

☻ Other legal approaches can be taken outside of the contract.

☻ The terms of the contract can only be enforced by the client or contractor taking action.

11. Law of the contract

Only English or Welsh law applies to the contract.

The elements covered in the Building Contract for a Home Owner/Occupier who has not appointed a consultant to oversee the work are comprehensive yet straightforward enough to cover the fundamental eventualities of a domestic building project going wrong. When implemented with common sense, these contracts protect both the client's and the contractor's interests. They remove any concerns from the client in particular, as they spell out what a competent contractor should already be providing.

The Building Contract for a Home Owner/Occupier who has appointed a consultant to oversee the work is presented in a pack/folder. It contains:

☻ Two building contracts:
- one for the builder

- one for the client.
- Two consultancy agreements:
  - one for the consultant
  - one for the client.
- Guidance notes

The Building Contracts are almost identical to those in the Building Contract for a Home Owner/Occupier who has not appointed a consultant to oversee the work. However, there are some minor differences in the contract of which the following are worth pointing out.

The consultant's role

- The consultant will act on the client's behalf to:
  - instruct the contractor on all issues
  - extend timescales if required
  - issue payment certificates.
- The consultant will issue certificates to client and contractor.
- The legal framework of the contract is between the client and contractor.

Certifying finished work and 'making good'

- When the work is inspected and satisfactorily complete, a certificate will be issued by the consultant showing the date of completion.
- A list of defects or faults will be issued for correction by the contractor (for which he is liable), no later than three months after completion.
- When the faults or defects have been rectified, a further certificate will be issued.

The issuing of certificates relates directly to payments and completed work. The consultant will not issue certificates unless the invoiced work has been checked and found to be satisfactory.

As with the building contract, the Consultancy Agreement is in two parts:

- Part 1 The Consultant's Services (alphabetised breakdown)
- Part 2 The Conditions (numbered breakdown)

Each part is broken down into separate sections, which cover the specifics of the agreement.

Part 1 – The Consultant's Services cover:

A. The services
B. Consultant's fees
C. Insurance
D. Disputes

A. The services

This section is set out in four stages that are fairly detailed but very easy to understand. They identify the services the consultant will provide. Rather than go into every detail, here is a short summary of each stage:

Stages 1, 2 and 3 are made up of a series of boxes that the client can tick to show which services, what advice or assistance will be required before any work starts on site. For example:

- **Stage 1**
  - Client's needs and budgets
  - Surveys that will be required
  - Sketches and guidance on costs and timescales
  - Preparation of reports
  - Advice on planning permission and procedures
  - Agreeing fees
  - Advice on the contract to use if the consultant is not providing services at Stage 4: After building work has started
- **Stage 2**
  - Producing drawings for the work
  - Preparing the specification
  - Applying for planning permission
  - Applying for building regulations
  - Advising and taking action where required on party wall issues
  - Co-ordination of other consultant's (appointed by client) work
- **Stage 3**
  - Explanation of the terms of contract and assistance with its completion
  - Obtaining quotes
  - Deciding on which contractor to use
  - Explanation of health & safety and environmental issues
- **Stage 4 – After building work has started**
  Stage 4 has a single tick box to indicate that the consultant should:
  - inspect the work
  - ensure the contract is being adhered to
  - advise on timescales and costs in relation to changes
  - give instruction to the contractor
  - check all invoices from the contractor
  - issue certificates (if satisfied) for payment
  - prepare lists (if necessary) of faults
  - issue finishing certificate.

B. Consultant's fees

- Shows the total of the consultant's fee and at what stages it will be paid.
- Shows agreed hourly rate for other services.

- Shows the costs that are not the consultant's responsibility, e.g.:
  - any local authority fees
  - party wall surveyors
  - other consultants.
- Changes to the services will be subject to condition B.3.

C. Insurance
- Shows the minimum amount of professional indemnity that the consultant holds.
- Confirms the consultant will provide proof of insurance if required.

D. Disputes
- The client or consultant can start court proceedings if a dispute arises between them.
- An adjudicator can be appointed to settle a dispute.
- The consultant will not appoint an organisation of which he is a member, in the event of a dispute.
- Both parties agree that costs of an adjudication will become part of the agreement.
- Organisations providing free details for costs, rules and procedures are listed.

Part 2 – The Conditions cover:
1. Consultant's responsibilities
2. Client's responsibilities
3. Changing the services
4. Paying the consultant's fees
5. Consultant's continuing responsibilities
6. Copyright
7. Bringing the agreement to an end
8. Other rights and remedies
9. Law of the contract

1. Consultant's responsibilities
- The consultant will:
  - act as the client's representative
  - carry out the work professionally
  - be fair in all undertakings
  - co-operate with other consultants
  - meet all legal obligations.
- The consultant will not without the client's permission:
  - sub-contract agreed services
  - allow the contractor to sub-contract the work.

2. Client's responsibilities
- The client will:
  - provide appropriate information to the consultant
  - consider the advice of the consultant
  - allow the consultant to deal with the contractor.

- The client will not:
  - interfere with the consultant's set duties
  - deal directly with the contractor.

3. Changing the services

Only the client can change the services.

- If changes increase the services, agreements must be made about fees before they are carried out.
- If the changes reduce the service, the consultant will reduce the fees accordingly.

4. Paying the consultant's fees

Fees will be payable 14 days from date of invoice.

5. Consultant's continuing responsibilities

The consultant will be responsible for the consequences of failing to keep to the agreement for six years after completion.

6. Copyright

Copyright of all material produced will remain that of the producer.

7. Bringing the agreement to an end

- The agreement can be ended with seven days' written notice by either party; no reason need be given.
- Once a contract is signed between client and contractor:
  - If the client makes a formal written complaint about the services not being carried out properly, the consultant must put matters right within seven days. The client can terminate the agreement if this is not done by issuing a further notice, which will take immediate effect.
  - If the client does not pay the consultant's fees or obstructs his duties, the consultant can make a formal written complaint. If the client does not put matters right the consultant can terminate the agreement by issuing a further notice, which will take immediate effect.
- If either party becomes insolvent (unable to meet debts) the contract will end unless other arrangements are made by a professional insolvency practitioner to allow the contract to continue.
- If the agreement ends, the client only has to pay for work carried out up to the date of the termination.

8. Other rights and remedies

- Claims can be made by either party for costs as a result of the terms of the agreement being broken by one or the other.
- Other legal approaches can be taken outside of the agreement.
- The terms of the agreement can only be enforced by the client or consultant taking action.

9. Law of the contract
Only English or Welsh law applies to the agreement.

As you can see from the detail in the contract and the agreement, most of the fundamental elements have been covered. Too many cases are brought to court that could easily have been avoided by putting in place a simple contract. Having a contract does not ensure that the work will be carried out on time or that the standards will be to your satisfaction, but it does provide a system to deal with problems such as these. I would advise anyone preparing to employ the services of a builder or contractor to consider seriously what safeguards are going to be put in place.

Let us look at the options in relation to different size projects:

You may not want to go to the lengths of implementing a formal off-the-shelf contract if you are going to employ a specialist contractor to carry out a specific task on a small project, such as:

- new windows
- central heating
- new electrical system
- new roof covering.

SAMPLE SELF-PREPARED CONTRACT

Client's name:
Client's address:
Project address:
Tel. No:
Contractor's name:
Contractor's address:
Tel. No:
Description of work: Replacement of all windows and external doors with type as agreed in showroom. All internal and external making good and remedial work to surrounding flowerbeds to be carried out by contractor and co-ordinate with existing. Contractor to refix curtain rails and runners.
Planning permission: Arranged by client.
Facilities: Client's washroom/toilet can be used provided it is kept clean.
Start date: --/--/--   **Finish date:** --/--/--
Working times: 8am–6pm
Guarantees: To be given to client.
Health & safety: All appropriate health & safety measures must be taken.
Security: Adequate security measures to be taken at all times.
Price and Payment: As terms and conditions of order with contractor/supplier.
Defects and retention: Contractor/supplier to be responsible for defective workmanship for a period of three months after installation in addition to that stated in guarantees. The client will withhold 5% of overall payment until the three months has expired.
Client signature:
Contractor signature:
Date:

In this case you should think about preparing a contract that states exactly what your requirements are. You could draw up a contract to ensure that your requirements are understood. The previous page shows a very basic example of a self-prepared contract that would need to be signed and given to both parties. Some contractors or suppliers may have their own terms of agreement, and if this is the case make sure that you read the terms and are fully satisfied with them before proceeding.

Where you are employing various contractors to carry out work, you may need to consider the impact of the work on the immediate surroundings and environment. Apart from those written in the specification you will need to spell out in the contract any terms that you would like to implement during the period of contract.

If you are preparing to employ the services of a single builder or contractor to under-take a project of reasonable size for a domestic property, you would be better off using a pre-formatted contract such as the Joint Contract Tribunal (JCT) Building Contract for a Home Owner/Occupier who has not appointed a consultant to oversee the work. This contract would be more suited to the following types of projects:

- small extensions
- small loft conversions
- part refurbishment
- small basement conversion
- demolition
- swimming pool.

Although the contract is for those who are not appointing a consultant, it does not prohibit you using a building consultant to check the work on your behalf. An independent adviser would have no contractual say in any disputes arising.

The type of project more suited to the (JCT) Building Contract for a Home Owner/Occupier who has appointed a consultant to oversee the work would be:

- building a new house
- large extensions
- large loft conversions
- total refurbishment
- large basement conversion
- demolition and building
- swimming pool and outbuildings
- technically difficult projects
- unusual projects
- large external works
- architectural landscaping.

Although on a large project the professional fees may amount to a considerable sum, consultants are usually trained, qualified and experienced in dealing with contractors. If you are planning to invest large amounts of money and are not fully confident in implementing the terms of a contract, serious thought must be given to the type of contract you will use.

Retention of payment and defect liability period

Whether you will be making stage payments or one final payment, it is advisable to hold back 5 per cent for a period of three months, or a mutually agreed period. This is known as *the defects liability period*. The reason for the retention of payment is to ensure that in the event of defective work, the contractor has an incentive to return and rectify any problems that he is responsible for. Even if the work requires remedial action before the agreed defects liability period has expired, and the contractor does attend to rectify the work, the 5 per cent should still be held until the expiry date. The final 5 per cent payment can then be made upon a final inspection with the contractor.

If a contractor has been requested to carry out work under the terms of an agreement and is not co-operating with your requests, it is advisable to put requests in writing. Give the contractor notice that if necessary you will have the remedial work carried out by another contractor, and use funds owed from the retention money to pay for the work. In the event of emergency work being required (as a result of a contractor's work being defective), the contractor would be expected to provide the emergency cover. In the event that the contractor did not provide sufficient cover, payment for the emergency work may be deducted from the retention, in order to have the work carried out by another contractor.

Chapter 4

Chapter 4

getting going

Financial arrangements

Before any project has started, it is very important that the builder and/or contractors you will be using are aware of the financial arrangements you will be implementing. This is not something that you should have a fixed idea about as all builders and contractors will have their own ideas on how they see the payments being made. However, you do need to have an agreed system for making payments where you are in control and where both parties are in agreement, with a clear understanding of the details.

As a general principle, it is inadvisable to pay for any work in advance. Only where goods need to be made off-site for the project (such as specialist joinery work) should you consider doing so. If you are asked to make advance payments for off-site work, ask for a breakdown and then you should only pay a deposit. Situations such as these are obviously subject to negotiation; and you should always obtain a receipt for any payments that you make.

In general, payments for work are made on the amount of work that has been carried out, not on the amount of time that has elapsed. For example:

- brickwork to damp-proof course (DPC)
- brickwork to roof level (plate)
- roof structure and covering
- external windows and doors
- first fix carpentry
- first fix mechanical and electrical (M + E)
- plastering
- second fix carpentry & M + E
- decorating.

123

The stages of a project can be seen easily by looking at the programme. This can be used as a guide to measure the amount of work carried out in relation to the timescale. Never pay for work that has not been carried out, even if the builder says he needs the money to pay for materials. It is likely that there is a cash flow problem and you could be risking your money to bail the company out on another project.

Insurance details

Whatever project you are planning, it is important to ensure that both your own and your contractor's insurance are sufficient. If you are planning a self-build project, self-build site insurance is an absolute necessity. In addition, structural warranty insurance may be required by your mortgage lenders, as part of their lending criteria, which should cover a period of no less than ten years. This insurance should protect your project against a defect in the design, workmanship and materials.

Apart from the peace of mind you gain with this type of product, if your circumstances change and you wish to sell the property within ten years after completion, you may find that your buyer's lender requests a structural warranty for the building before they lend them any money.

Products such as these have an advantage over an architect's certificate. This is really only an extension of the architect's Professional Indemnity insurance, requiring you to prove negligence of design (or other factors) causing defects at a later stage.

Summary of Cover for self-build site insurance would include:

- the cost of complete or partial rebuilding or rectifying work to the new housing unit which has been affected by major damage caused by a defect in the design, workmanship or materials
- the necessary and reasonable costs in repairing, replacing or rectifying any part of the waterproof envelope of the housing unit as a result of ingress of water caused by a defect in the design, workmanship, materials or components of the weatherproofing elements
- the cost of making good any defect in the design materials or workmanship in the drainage system, which was newly constructed by the builder
- the cost of repairing or making good any defects in the chimneys and flues of the new housing unit causing an imminent danger to the health & safety of occupants.

Site insurance is of fundamental importance when investing your money into property, especially where the sums are substantial. Whether your project is a conversion, renovation, extension or a new build, you will invariably need comprehensive protection for both the building and the rest of the site. You must protect the existing structure (if there is one) as well as all the new work against losses such as fire, theft, flood, storm damage, vandalism and accidental damage. Do not rely solely on your standard building insurance to cover the cost of any loss. Frequently they will not pay out on a loss involving a property under development.

You can find out on-line by visiting: **www.selfbuildzone.com** This website explains exactly the type of insurance cover that you will need, and also provides quotes.

When it comes to insurance for builders and contractors, there are many types covering all the different trades within the construction industry. The most common type of policy for builders is the 'contractor's all risk insurance policy'. This covers:

- Public liability insurance:
  - protection from any liabilities from a third party (other than their employees), bodily injury or damage to property that may occur during the normal operation of their business.
- Builder's Employer's liability insurance:
  - builder's staff, including part-time staff, trainees or sub-contractors that they employ.
- Builder's Products liability insurance:
  - liabilities from a third party's bodily injury (other than their employees), or damage to property that may occur from products sold or supplied by the contractor.
- Builder's tools, plant and equipment insurance:
  - against loss, damage or destruction on- or off-site.

Insurance cover cannot be left to chance. It is therefore in your interests to satisfy yourself that sufficient insurance cover is provided by you or your chosen contractor(s). Your own insurance company may be able to advise you on the type of insurance cover that should be in place before the project starts.

Party Wall Issues – The Party Wall etc. Act 1996

If you are intending to carry out work on a party wall, you may need to consult a Party Wall Surveyor. They will ensure that all regulations under the Party Wall etc. Act 1996 are adhered to. However, this will depend on the nature and complexity of the work. The Act may affect someone who either wishes to carry out work covered by the Act, i.e. the 'Building Owner', or someone who receives notification under the Act of proposed adjacent work, i.e. the 'Adjoining Owner'. Under the Act, the term 'owner' means the persons or body holding the freehold title, or holding the leasehold title for a period exceeding one year. There are other circumstances where the Act would come into effect, for example where a person is under contract to purchase such a title, or is entitled to receive rents from the property. Therefore, there may be more than one set of 'owners' of a single property.

Some minor works to a party wall may not necessarily obligate you to inform adjoining owners. However, if the planned work has potential to cause structural damage or affect the strength of the wall, it would be advisable to seek the advice of a qualified building professional.

Work that would be considered as minor might include:

- drilling into a party wall to fix wall units, shelves or pictures
- cutting into a party wall to add or modify electrical recessed wiring or sockets
- removing existing plaster in order to re-plaster.

The Act provides a building owner who wishes to carry out major work to an existing party wall with additional rights, going beyond ordinary common law rights. The most commonly used rights are:

- to cut into a wall to take the bearing of a beam, for example for a loft conversion
- to insert a damp-proof course all the way through the wall
- to raise the height of the wall and/or increase the thickness of the party wall
- to demolish and rebuild the party wall
- to underpin the whole thickness of a party wall
- to protect two adjoining walls by putting a flashing from the higher over the lower, even where this requires cutting into an adjoining owner's independent building.

If you intend to carry out major works to a party wall, you must inform all adjoining owners at least two months before the planned starting date. The adjoining owner may agree to allow the work to start before the two months have elapsed, but they are under no obligation to, even when agreement on the works is reached. A notice is only valid for one year, so do not serve it too soon before you wish to start. You must not cut into your own side of the wall to carry out any part of a major element of work without informing or giving notice to the adjoining owners about your intentions.

Remember that reaching agreement with the adjoining owner or owners does not remove the possible need to apply for planning permission, or to comply with building regulations procedures. Conversely, obtaining planning permission or complying with the building regulations does not remove the need (where applicable) to comply with the Act. If you intend carrying out any type of building work involving the following:

- work on an existing wall or structure shared with an adjoining owner,
- building a free-standing wall or a wall of a building up to or astride the boundary with a neighbouring property,
- excavating near a neighbouring building,

you must ascertain whether the work falls within the Act. If it does, you are legally obliged to notify all adjoining owners.

The Act provides a legal framework for preventing and resolving disputes in relation to party walls, boundary walls and excavations near neighbouring buildings. The Act is based on some tried and tested provisions of the London Building Acts. If you are intending to carry out work (anywhere in England and Wales) of the type described in the Act, you must give notice to adjoining owners of your intentions. If the intended work is to an existing party wall a notice must be given, even if the work will not extend beyond the centre line of a party wall.

Where there is no written consent or agreement, the Act provides for the resolution of disputes. The Act covers:

- various types of work that are going to be carried out directly to an existing party wall or structure
- new building at or astride the boundary line between properties
- excavation within 3 or 6 metres of a neighbouring building or structure, depending on the depth of the proposed hole or foundations.

The Act recognises two main types of party wall. These are:

Party wall type (A)

- A wall is a 'party wall' if it stands astride the boundary of land belonging to two or more different owners.
  - Such a wall is part of one building, or separates two or more buildings.
- A 'party fence wall'
  - A wall is a 'party fence wall' if it is not part of a building, and stands astride the boundary line between lands of different owners and is used to separate those lands, for example a garden wall (this does not include wooden fences or garden fences).

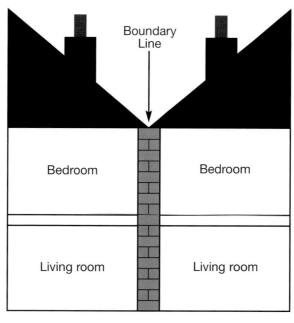

Party wall type A

Party wall type (B)

- A wall is also a 'party wall' if it stands completely on one owner's land, but is used by two or more owners to separate their buildings.
  - An example of this is where one person has originally built the wall, and another has abutted their building against it without constructing their own wall. Only the part of the wall that does the separating is 'party'. The sections on either side or above are not recognised as 'party'.

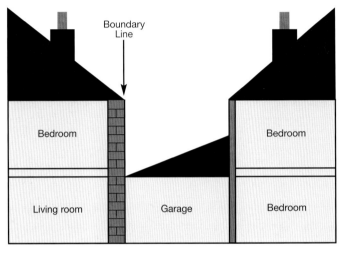

Party wall type B

The Act also uses the expression 'party structure'. This is a wider term to describe a wall, floor partition or other structure separating buildings or parts of buildings approached by separate staircases or entrances, e.g. flats or multiple properties within one building.

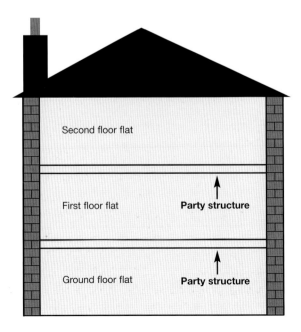

Issuing a notice

You may deliver the notice to the adjoining owner(s) in person, or send it by post. Where the neighbouring property is empty or the owner is not known, you may address the notice to 'The Owner', adding the address of the premises, and fix it to a conspicuous part of the premises. You do not need to tell the local authority about your notice.

While there is no official form for giving notice under the Act, your notice must include the following details:

- your own name and address (joint owners must all be named)
- the full address of the building to be worked on
- a full description of the proposed work (it may be helpful to include plans)
- when you propose to start (this must not be before the relevant notice period has elapsed, e.g. two months)

The notice should be dated and made clear that it is a notice under the 'provisions' of the Party Wall etc. Act 1996.

> **The Act does not** include enforcement procedures for failing to inform or serve a notice on an adjoining owner. However, if you did start work without having first given notice, the adjoining owners have the right to seek to stop your work through a court injunction or seek other legal redress. An adjoining owner cannot prevent someone from exercising the rights given to them by the Act. However, they may be able to influence the methods and timing of the proposed work.

Adjoining Property Owners may:

- give consent in writing
- object to the proposed works, in writing
- do nothing.

If a period of 14 days has elapsed since the service of your notice and the adjoining owner has not responded, a dispute is regarded as having arisen. You would obviously need to try and find out if there is a dispute, and if so to try settling any point of difference by friendly discussion. Agreements must always be put in writing.

If you cannot reach an agreement with the adjoining owners, it is advisable to discuss the next course of action. This would be to appoint what the Act calls an 'Agreed Surveyor', to draw up an 'Award'. The Agreed Surveyor needs to be totally impartial to the project and have good knowledge of the construction industry, and in particular should have experience in administering the Act. It would be advisable to contact a qualified building professional with some experience or knowledge of party wall matters.

If the appointment of one surveyor does not satisfy the adjoining owner, you can each appoint a surveyor to draw up the award together. The two appointed surveyors should have a good understanding of the Act. However, a third surveyor would be called

NOTICE OF ADJACENT EXCAVATION

Party Wall Etc. Act 1996 section 6

To: Mr Smith   (Adjoining owner)
of   123 Peter Road
           Leigh, Essex

We   Mr Jones.   (Building owner)
of   121 Peter Road
           Leigh, Essex

as owners of   121 Peter Road, Leigh, Essex

which adjoins your premises known as 123 Peter Road Leigh Essex

HERERBY SERVE YOU WITH NOTICE THAT IN ACCORDANCE WITH OUR RIGHTS

Under section 6(1)
It is intended to build within 3 metres of your building and to a lower level than the bottom of your foundations, by carrying out the works detailed below, after the expiration of one month from the service of this notice.
and
*Under section 6(2)
it is intended to build within 6 metres of your building and to a depth as defined in the Act, by carrying out the works detailed below, after the expiration of one month from the service of this notice.

IT IS NOT PROPOSED TO UNDERPIN OR OTHERWISE STRENGTHEN IN ORDER TO SAFEGUARD THE FOUNDATIONS OF YOUR BUILDING.

The accompanying plans and sections show the site, the foundation type and indicative depth:
407.01.00
407.01.01
407.01.02
L3224 SL(20)010
Ordnance Survey Identification Plan

The intended works are:

Demolish the existing building and construct a new 3-Storey Detached House with basement under on raft foundations, excavating for foundations approximately 5m in depth.

It is intended to commence works as soon as notice has run or earlier by agreement.

Under section 6(7), if you do not consent to the works within **14 days** you are deemed to have dissented and a dispute is deemed to have arisen. In such case section 10 of the Act requires that both parties should concur in the appointment of a surveyor or should each appoint one surveyor and in those circumstances
We would appoint

Mr A N Other
of The Livemore Partnership, Broadway House, 74-76 Broadway, Leigh-on-Sea, Essex SS9 1AE.

Signed
*Authorised to sign   26 September 2005
(Date)
On behalf of Mr Jones   (Building owner)

MC/6219

* delete as appropriate

An example of a notice of proposed adjacent excavation

in if the two appointed surveyors could not agree. In all cases, surveyors appointed under the dispute resolution procedure of the Act must consider the interests and rights of both owners, and draw up an impartial award.

A person who receives notice about proposed work to a party wall may, within one month, give a written counter-notice setting out what additional or modified work they would like to see carried out for their benefit. The adjoining owner should notify you within 14 days of your notice if they intend to send a counter-notice. If you receive a counter-notice, you must respond to it within 14 days, otherwise a dispute is regarded to have arisen.

> **If you obtain** consent you are still obligated under the Act to avoid unnecessary inconvenience, and provide temporary protection for adjacent buildings and property where necessary. The notice of consent is confirmation that the adjoining owner does not dispute or object to the proposed work. Should a problem arise at a later date such as damage caused by the work, you should try to resolve things with the adjoining owner. If this causes further dispute, the surveyor(s) should be called in.
>
> Where separate surveyors are appointed, the surveyors must liaise with their appointing owners and put forward the preferred outcome. However, the surveyors do not act as supporters of the respective owners. They must always act within their statutory jurisdiction, and jointly prepare a fair and impartial award.

The award

The surveyor(s) will prepare an 'award' (also known as a 'party wall award'). This document:

- sets out the work that will be carried out
- states when and how the work is to be carried out
- specifies additional work required, e.g. protection work to prevent damage
- contains a record (condition schedule) of the adjoining property before the work begins in order to attribute any damage caused to adjoining land or buildings
- allows access for the surveyor(s) to inspect the ongoing works in order to ensure full compliance with the award.

> **It is advisable** to keep a copy of the award with your property deeds when the work is completed.

Payment

The surveyor(s) will decide who pays the fees for drawing up the award, and for carrying out the necessary checks to ensure that the work has been carried out in accordance with it. If the work is totally for your benefit, it is likely that all costs will be paid by you. Your agreement with the adjoining owner for payment of the work will be covered in the award. The general principle in the Act is that the person who initiated the work pays for it if the work is solely for his benefit. However, there are cases where the adjoining owner may pay part of the cost, for example:

- if work to a party wall is required because of defects or lack of repair for which the adjoining owner may be in full or part responsible
- where an adjoining owner requests additional work to be done for his benefit
- where the dispute resolution procedure is called upon, the award may deal with apportionment of the costs of the work
- the dispute procedure may be used specifically to resolve the question of costs if this is the only matter in dispute.

Appeals

The award is final and binding unless it is appealed against and amended by the court. Each owner has 14 days to appeal to the county court against an award. Appealing against an award should be considered at length and it would be advisable to seek legal advice. An appeal should only be made to the county court if an owner believes that the surveyor(s) has made fundamentally wrong decisions.

Lack of co-operation

If an adjoining owner is disputing the work but refuses to appoint a surveyor, under the dispute resolution procedure you are entitled to appoint a second surveyor on their behalf so the procedure can go ahead. In these circumstances, you will not be able to appoint an 'agreed surveyor'; this will be the responsibility of your own surveyor who will appoint a second surveyor on behalf of the adjoining owner.

Access to adjoining properties

Under the Act, an adjoining owner and/or occupier must allow access to your workmen, surveyor or architect when necessary, provided that it is in regard to carrying out works in pursuance of the Act. They must also allow access to appointed surveyors as part of the dispute resolution procedure. However, you must give the adjoining owner and occupier a period of usually 14 days' notice of your intention to exercise these rights of entry. It is an offence to refuse entry to or obstruct someone who is entitled to enter premises under the Act. The offender can be prosecuted in the magistrate's court, if they know or have reason to believe that the person is entitled to be there. If the adjoining property is empty or unoccupied, your workmen and your own surveyor or architect may enter the premises only after following the correct procedures. They will need to be accompanied by a police officer.

If there are other works that are not covered under the statutory rights of access, it is reasonable to expect that amicable agreements are reached to ensure that the work is carried out to a high standard. You should discuss access for other works with your neighbour. It is often in the best interests of the adjoining owner to allow access voluntarily to build a wall, or carry out works that will enhance the quality of finish to the side of the wall that they will see.

Adjoining owner's rights

Adjoining owners have the right to:

- appoint a surveyor to resolve a dispute
- request that reasonable measures are taken to protect their property from possible damage
- be protected from any unnecessary inconvenience
- compensation for any loss or damage caused by relevant works
- request security of expenses before commencement of work to minimise the risk of being left in difficulties if you stop work at an inconvenient stage.

The Act states that the building owner must not cause unnecessary inconvenience. This means inconvenience that is considered over and above that which will obviously occur with the type of work undertaken. The Building Owner must provide the appropriate temporary protection for adjacent buildings and property where necessary. The building owner is also responsible for making good any damage that is caused by the works, or must make payment in lieu of making good if requested by the adjoining owner.

Building a party wall or party fence wall astride the boundary line

If you are planning to build a party wall or party fence wall astride the boundary line, you must inform the adjoining owner by serving a notice as described earlier. However, there is no right to build astride the boundary without your neighbour's agreement in writing. If you plan to build a wall completely on your own land but up against the boundary line, you must also inform the adjoining owner by serving a notice

You will need to serve the notice at least one month before the planned start date for building the wall, and the notice is only valid for a year. If the adjoining owner agrees within 14 days for the building of the wall to go ahead, the work (as agreed) may proceed. The cost of building the wall may be shared between you and the adjoining owner if the benefits and use of that wall will be shared. All agreements must be in writing and should record details and specification of the wall, the location, and the allocation of costs and any other agreed conditions. If the adjoining owner does not agree (in writing within 14 days), to the proposed new wall astride the boundary line, you may only build the wall wholly on your own land, and wholly at your own expense. However, you do have the right to place appropriate footings for the new wall under your neighbour's land, providing that compensation for any damage caused by building the wall or laying the foundations is addressed by you.

If there is a disagreement about any work, the same procedures will apply (as stated earlier), with the appointment of an 'agreed surveyor'. The surveyor(s) can assist the owners in reaching an agreement, but they cannot make decisions if the boundary location is in dispute. The Act does not contain any provision to settle a boundary line dispute. Disputes such as these can be resolved through the courts or through alternative dispute resolution procedures, for example arbitration, mediation or decision by an independent expert. This may be simpler, quicker and cheaper than court proceedings.

Excavating near neighbouring buildings

As before, you must inform the adjoining owner or owners by serving a notice if you plan to:

- excavate, or excavate and construct foundations for a new building or structure, within 3 metres of a neighbouring owner's building or structure, where that work will go deeper than the neighbour's foundations
- excavate, or excavate for and construct foundations for a new building or structure, within 6 metres of a neighbouring owner's building or structure, where that work will cut a line drawn downwards at 45° from the bottom of the neighbour's foundations.

> **'Adjoining owners'** may include your next-but-one neighbour if they have foundations within 6 metres of your excavation.

The notice must state whether you propose to carry out any strengthening or support work to the foundations of the building or structure belonging to the adjoining owner. You must provide plans and sections of the work showing the location and depth of the proposed excavation or foundation. The location of any proposed building must also accompany the notice. You must give notice at least one month before the planned start date for the excavation.

In the event of damage caused to adjoining owner's property, you will be legally responsible for putting it right, even if the damage is caused by his contractor.

Needless to say if the adjoining owner wants to undertake building work that affects the party wall, they owe the same duty to you under the Party Wall etc. Act 1996, and the same procedures will apply.

Condition schedule

A condition schedule is usually associated with the letting of properties whereby the condition of walls, carpets, furniture, etc., is documented at the beginning of the let and then inspected for damage at the end. If there is any damage other than that recorded and that cannot be classed as reasonable wear and tear, the tenant would be responsible for the cost of remedial work. If you are planning to have work carried out to a property where the work involves builders or contractors having regular access through the property (or for example, working in areas where the finishes are of a high quality), the schedule would help to identify the level of temporary protection required.

Condition schedules are not restricted to internal use. If you are only having external work carried out, you could draw up a schedule to identify existing defects and specify precautions that you would like to have put in place. After the condition schedule is prepared, it should be dated and signed by both parties to ensure that the descriptions are a true reflection of the conditions found at the time. Once all the work is complete and before final payments are made, a further check of the items on the schedule should be carried out. Where damage has been caused during the new work, the contractor should either repair the damage or make proposals to have it repaired if it is not within his scope, at his own cost.

Environmental considerations

Most homeowners will not be aware of the legal responsibilities placed on the designers of houses and structures generally and may question some of the details of the specification. As stated previously, architects and engineers have to comply with regulations. However, it is useful to understand what is covered under environmental considerations.

There are particular insulation factors that must be met regarding all building materials; these are known as 'U' values. 'U' values (the overall co-efficient of heat transmission) indicate the heat-flow through materials, and although values vary depending upon the type of material, they will combine to give an overall value. It is important that the performance rating of the building complies with the building regulations. An example of this is the structure of a traditional house wall that would include the following materials:

- Outer skin of brickwork
- Cavity wall insulation
- Blockwork or studwork
- Plaster or plasterboard

Each of these elements will have different 'U' values, and each will have to meet a specific rating that is dependent upon the use of the building. This is not an area that need concern you in great technical detail, but those with a more 'green' conscience may find the subject interesting.

The building or refurbishment of a house or extension will obviously require natural material resources to be used. We are all responsible for ensuring that 'best practice' is incorporated in our design. This not only requires us to use materials that meet regulations for insulation and structural compliance, but also that we are being considerate environmentally.

In *The Housebuilder's Bible* by Mark Brinkley (6th edition, Ovolo Publishing, 2004) there is specific reference to the extraction of aggregates, and the effect of using wood-based products, in particular tropical hardwoods. We are becoming more conscious of our environment and surroundings, and of those who live in areas where these materials are extracted, and consequently look more and more to alternatives to avoid further undermining of the resources available to us. However, it should be borne in mind that alternative products may cost a little more.

As discussed in *Scheme concept and initial ideas* (Chapter 1), it is important to keep your ideas achievable, real and in keeping with the local surroundings. Whatever project you have in mind it is important, in addition, to be environmentally aware.

Trees

It is not widely appreciated that many trees are protected by tree preservation orders, or TPOs. This means that, in general, you need the local council's consent to prune or fell them. If you live in a conservation area you need to be especially aware of this. If you are in any doubt about trees within your local environment that may be affected by your project, ask the council for a copy of their free leaflet: *Protected Trees: a guide to tree preservation procedures*. You can be fined for any unlawful work to trees which are protected by a TPO.

Tree Preservation Orders

A tree preservation order is an order made by a local planning authority (LPA) in respect of trees or woodlands. The principal effect of a TPO is to prohibit:

- cutting down
- uprooting
- topping
- lopping
- wilful damage
- wilful destruction.

As trees are now becoming a more important feature of our local 'built' environment, it is important that we recognise their significance to us all. Once a tree is removed it is removed for good and, no matter what the fine, it will take years for another one to grow enough to replace it!

LPA's consent may be required to carry out any work to, or for the complete removal of, a tree. The cutting of roots (although not expressly covered in the bullet points above), is potentially damaging for several reasons. These include the danger that the tree can become unstable in high winds, and it may pose specific dangers to people and property. It is therefore the view of the Secretary of State that in many cases the LPA's consent must first be obtained.

Law

The law on TPOs is in Part VIII of: *The Town and Country Planning Act 1990 ('the Act')* and in *The Town and Country Planning (Trees) Regulations 1999 ('the 1999 Regulations')* which came into force on 2nd August 1999.

The dictionary defines a tree as a perennial plant with a self-supporting woody main stem, usually developing woody branches at some distance from the ground and growing to a considerable height and size. But for the purposes of the TPO legislation, the High Court has held that a 'tree' is anything that would ordinarily be called a tree.

SECTION 201 DIRECTIONS

Regulations for LPAs concerning TPOs are outlined in detail under Section 201: *Tree Preservation Orders – a guide to the law and good practice*. These guidelines can be obtained from the Office of the Deputy Prime Minister (www.odpm.gov.uk).

LOCAL PLANNING AUTHORITIES (LPA)

The power to make a TPO is exercised by the LPA. In England, the LPA is the district, borough or unitary council. A county council may make a TPO, but only:

- in connection with the grant of planning permission
- on land which is not wholly within the area of a single district council
- on land in which the county council hold an interest
- on land in a national park.

Special arrangements apply in the following areas:

- in national parks where the National Park Authority are responsible for TPO functions, concurrently with the district, borough or unitary council
- in the Norfolk and Suffolk Broads; the Broads Authority are responsible for TPO functions concurrently with the district, borough or unitary council
- in enterprise zones, urban development areas and housing action trust areas, where the Enterprise Authority, Urban Development Corporation or Housing Action Trust are the sole LPA.

When they make a TPO, these authorities are advised to copy it for information to the tree officer of the district, borough or unitary council, within whose area the trees or woodlands are situated.

As you can imagine, the nature of any project (no matter how large or small) may make immediate neighbours or interested third parties keen to know to how it will affect them. When it comes to trees (even if they are on your land) there are people who would go to extreme measures to protect them. It is important that if you are planning to cut down a tree (for your own peace of mind), to ensure that you are acting within the law.

If you were to assume that a tree within your own property could be cut down and you failed to clarify it, you could find that it does have a TPO. Your project could be held up until the legal situation is confirmed, or you could even be fined if you acted in a way that proved you were negligent.

The Secretary of State's powers

The Secretary of State for the Environment, Transport and the Regions ('The Secretary of State') has the power to make a TPO. In considering requests to make a TPO, the Secretary of State will have regard to all representations submitted to him. However, it is likely that he would use his power only in exceptional circumstances, where issues of more than local significance are raised, and then only after consultation with the LPA in whose area the trees or woodlands are located.

Trees and woodlands

A TPO protects trees and woodlands. The term 'tree' is not defined in the Act, nor does the Act limit the application of TPOs to trees of a minimum size. Fruit trees, for example, may be included in a TPO, provided it is in the interests of amenity or enhancement of a public area.

If your proposed project or development involves building in or on a wooded area, you need to ensure that an existing TPO will not restrict your plans. When buying land for development this is one area that needs to be fully understood.

There is no definition of the term 'woodland' in the Town and Country Planning Act. In the Secretary of State's view, trees which are planted or grow naturally within the woodland area after the TPO is made are also protected by the TPO. This is because the purpose of the TPO is to safeguard the woodland unit as a whole and this depends on regeneration or new planting. As far as the TPO is concerned, only the cutting down, destruction or carrying out of work on trees within the woodland area is prohibited. Whether or not seedlings, for example, are 'trees' for the purposes of the Act, would be a matter for the courts to decide within the circumstances of a particular case.

Hedges

A TPO may only be used to protect trees and cannot be applied to bushes or shrubs. However, according to the Secretary of State, a TPO may be made to protect trees in hedges, or an old hedge which has become a line of trees of a reasonable height, and is not subject to hedgerow management. Separate legislation is in place to regulate the removal of hedgerows.

Local authority land

LPAs may make TPOs in respect of their own trees, or trees under their control. They may sometimes acquire land which is already the subject of a TPO. If the LPA (i.e. any department of the council as a whole and not just their planning department) propose to cut down or carry out work on protected trees, they may grant themselves consent. There would normally be good reasons for the cutting down of trees on local authority land, such as the effect that a tree may have on underground services or the age of the tree (if a tree is so old as to become unstable, it may also be in danger of falling, causing damage to persons or property).

The power to make a TPO

LPAs may make a TPO if in their opinion it appears to be in the interests of 'amenity' or the local environment to make provision for the preservation of trees or woodlands in their area.

Amenity: the Town and Country Planning Act does not define 'amenity', nor does it prescribe the circumstances in which it is in the interests of amenity to make a TPO. According to the Secretary of State, TPOs should be used to protect selected trees and woodlands if their removal would have a significant impact on the local environment and its enjoyment by the public. LPAs should be able to show that a reasonable degree of public benefit would accrue before TPOs are made or confirmed. The trees (or at least part of them), should therefore usually be visible from a public place, such as a road or footpath. In exceptional circumstances, the inclusion of other trees may be justified. The benefit may be present or future; trees may be worthy of preservation for their intrinsic beauty or for their contribution to the landscape, or because they serve to screen an eyesore or future development. The value of trees may be enhanced by their scarcity, and the value of a group of trees or woodland may only be collective. Other factors (such as the importance as a habitat for wildlife) may be taken into account, which alone would not be sufficient to warrant a TPO. According to the Secretary of State, it would be inappropriate to make a TPO in respect of a tree which is dead, dying or dangerous.

LPAs should be able to explain to landowners why their trees or woodlands have been protected by a TPO. They are advised to develop ways of assessing the 'amenity value' of trees in a structured and consistent way, taking into account the following key criteria:

- Visibility: the extent to which the trees or woodlands can be seen by the general public will inform the LPA's assessment of whether their impact on the local environment is significant. If they cannot be seen or are just barely visible from a public place, a TPO might only be justified in exceptional circumstances.
- Individual impact: the mere fact that a tree is publicly visible will not in itself be sufficient to warrant a TPO. The LPA should also assess the tree's particular importance by reference to its size and form, its future potential as an amenity,

139

taking into account any special factors such as its rarity, value as a screen, or contribution to the character or appearance of a conservation area.
- ◑ In relation to a group of trees or woodland, an assessment should be made as to its collective impact.
- ◑ Wider impact: the significance of the trees in their local surroundings should also be assessed, taking into account how suitable they are for their particular setting, as well as the presence of other trees in the vicinity.

EXPEDIENCY – CONVENIENT OR USEFUL

Although a tree may merit protection on amenity grounds, it may not be expedient to make it the subject of a TPO. For example, it is unlikely to be expedient to make a TPO in respect of trees which are under good arboricultural or silvicultural management. However, it may be expedient to make a TPO if the LPA believes there is a risk of the tree being cut down or pruned in ways which would have a significant impact on the amenity of the area. It is not necessary for the risk to be immediate. In some cases the LPA may believe that certain trees are at risk generally from development pressures. The LPA may have some other reason to believe that trees are at risk; changes in property ownership and intentions to fell trees are not always known in advance, and so the protection of selected trees by a precautionary TPO might sometimes be considered appropriate.

Making the TPO – internal procedures

TPOs are often made at a time when trees may soon be cut down or destroyed (which may be because a member of the public or somebody who has a specific interest has reported the imminent actions of another person), and therefore many LPAs find it convenient to delegate the function of making a TPO to an officer or officers of the council, and to put in place arrangements to act at short notice during and outside normal office hours. Officers acting on behalf of the council often retain the function of confirming TPOs, particularly where objections or representations have to be considered.

Site visit

Before making a TPO, the LPA officer will visit the site of the tree or trees in question and decide whether or not a TPO is justified. Any person duly authorised in writing by the LPA may enter land for the purpose of surveying it in connection with making or confirming a TPO, although the LPA may (in certain circumstances) decide to carry out the visit without entering the land. They may feel that the risk of felling justifies the making of a TPO before they have been able to assess fully the amenity value of the tree. However, this should not prevent them from making a preliminary judgement on whether a TPO would appear to be justified on amenity grounds, nor from making a more considered assessment before the TPO is confirmed.

Preparing the TPO – some considerations

A TPO must be in the form (or substantially in the form) of the 'model order' form of TPO included in the 1999 Regulations. The trees or woodlands to be protected must be specified in the 1st Schedule of the TPO, and their location shown on a map, included in, or annexed to, the TPO. The scale of the map (ideally an up-to-date Ordnance Survey map) must be sufficient to give a clear indication of the position of the trees or woodlands – 1:1250 will usually be sufficient for trees or groups of trees; 1:2500 will usually be sufficient for woodlands.

The Model Order provides that trees may be specified:

- individually; e.g. each tree: T1, T2, etc., circled in black on the map
- by reference to an area; e.g. the boundary of each area A1, A2, etc., indicated on the map by a dotted black line
- in groups; e.g. each group G1, G2, etc., shown within a broken black line
- as woodlands; e.g. the boundary of each woodland W1, W2, etc., indicated by a continuous black line.

Any combination of these four categories may be used in a single TPO.

During the site visit, the LPA officer will gather sufficient information to draw up the TPO. The LPA officer will accurately (subject to access) record the number and species of the individual trees or groups of trees to be included in the TPO and their location. In relation to areas of trees or woodlands, it is not necessary for the purposes of the TPO to record the number of trees, and a general description of species should be sufficient. However, it is important to gather enough information to be able to pinpoint the boundaries of the areas or woodlands in question on the map.

The LPA officer may also wish to record other information during the site visit for future reference, such as the present use of the land, the trees' importance as a wildlife habitat and trees which are not to be included in the TPO. If resources permit, the LPA may also find it useful to take photographs of the trees and their surroundings.

Individual and groups of trees

If trees merit protection in their own right, they will be specified as individual trees in the TPO. In general terms, the group category should be used for trees whose overall impact and quality merit protection. The intention of the group classification is not simply to protect trees which have individual merit and happen to be standing close to one another, but for their merit as a group.

Woodlands: woodland boundaries should be indicated on the map as accurately as possible, making use of any natural landscape features or property boundaries. This will avoid any future uncertainty if trees close to the boundary are removed. Use of the woodland classification is unlikely to be appropriate in gardens.

Completing the TPO

When completing a TPO, the LPA should ensure that they:

- include the title of the TPO
- include the name of the LPA
- insert the date on which it is to take effect
- apply the TPO to any trees which are to be planted pursuant to a planning condition if necessary
- complete the TPO and the map in a way that does not give rise to uncertainty
- enter the word 'NONE' against any of the classifications in the *1st Schedule* which are not used
- check that the TPO is complete, including both parts of the *2nd Schedule* and the map
- ensure the TPO is signed and dated.

Procedure when the TPO is made

On making a TPO, the LPA must serve on the owner and occupier of the land affected by the TPO:

- a copy of the TPO
- a notice ('Regulation 3 notice') stating:
  - the LPA's reasons for making the TPO
  - that objections or other representations about any of the trees or woodlands specified in the TPO may be made to the LPA
  - the date, being at least 28 days after the date of the Regulation 3 notice, by which any such objections or representations must be received by the LPA
  - the effect of the Section 201 direction if one has been included in the TPO.

The LPA must also make a copy of the TPO available for public inspection at the offices of the LPA.

The Regulation 3 notice is an important document. It may be the first document the owner or occupier sees concerning TPOs other than the TPO itself, which is long and complicated. The LPA is therefore advised to consider attaching to the notice some general information about TPOs, such as a copy of the department's explanatory leaflet: *Protected Trees: A Guide to Tree Preservation Procedures*. The LPA should also explain briefly the procedures that lead up to their decision on whether to confirm the TPO, as well as the name and telephone number of an officer who can give advice or answer any queries about the TPO.

Objections and representations

Objections and representations can be made on any grounds, for example:

- challenging the LPA's view that it is expedient in the interests of amenity to make a TPO

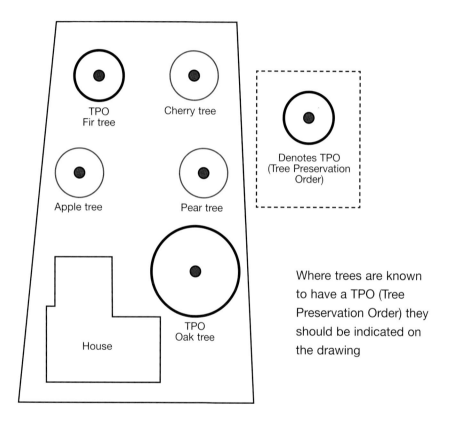

- claiming that a tree included in the TPO is dead, dying or dangerous
- claiming that a tree is causing damage to property
- pointing out errors in the TPO or uncertainties in respect of the trees which are supposed to be protected by it
- claiming that the LPA have not followed the procedural requirements of the regulations.

The LPA are required to take into account any objections or representations that have been made before deciding whether to confirm the TPO.

TPO
Fir tree

Cherry tree

Apple tree

Pear tree

Denotes TPO
(Tree Preservation
Order)

TPO
Oak tree

House

Where trees are known to have a TPO (Tree Preservation Order) they should be indicated on the drawing

Bar charts and programmes

Bar charts (or grids) have many applications which are used extensively throughout the construction industry and many other industries. It is the most straightforward way of seeing at a glance exactly what is required or planned. Even if you think you don't have enough information to make a chart or plan, don't be deterred from starting to put one together. You'll be surprised how just the mere fact of setting out your format for any given plan will prompt you into thinking about what needs to be included.

As you have already seen in Chapters 1 and 2, there are many uses for the grid system before we even touch on the bar charts for controlling the project. The following are just some of the other uses which will help to plan and control different aspects of the project:

- Initial planning stages
- Main programme of works – long term
- Short term – for periods of high activity
- Material ordering and delivery dates
- Payment schedules

When you start using these methods of controlling and monitoring progress, you will begin to realise that there are many other areas in which they will be useful. It is advisable to use these methods for planning and recording aspects of the project, no matter what stage you are at. If you do not have a system to work to, the information can become confusing or lost in the mountain of paperwork that is generated. These grid systems and bar charts will help you to keep the information concise and easy to monitor. Obviously there will be a great deal of information that needs to be filed. However, the idea of these systems is the easy retrieval of basic information or facts without the need to trawl through all of the documentation.

With any type of project, a timetable of the work needs to be established, so that the individual trades can carry out their work without clashing. Continuity of work is the key to completing the work in the allotted time. Whether you are using a builder or individual contractors, if you do not have a guide to when each element will be finished or when you expect certain trades to arrive, the continuity will suffer. This can have an adverse effect on the timescale and cost.

Bar chart programmes are by far the easiest to understand and can be produced by hand, on graph paper or on a computer, using a spreadsheet application. Bar chart programmes have many uses and can be formatted to suit your particular preference. In order to familiarise yourself, it is a good idea to use them from the outset of the project, from the very first planning stages through to the final details.

The individual elements will be monitored very easily using bar charts, which will help to reduce the chance of timescales being exceeded. All the information you need will be right there at your fingertips.

Initial planning stages

- Decide on requirements
- Arrange finance, if required
- Source professionals, architect, engineer, etc.
- Produce rough sketches
- Obtain outline drawings for planning permission
- Apply for planning permission
- Prepare specification of work
- List material quantities and suppliers

- Get contractors' quotes
- Receive planning permission
- Negotiate with chosen contractor(s)
- Start the project

The above is a sample of elements that you may need to consider. However, as every project is unique, there may be much more that you need to include, such as investigating particular service providers' timescales for upgrading or supplying new utility connections.

As you can see from this sample, there need to be defined timescales for everyone involved in the project to be aware of in order for you to plan your project start date. The fact that you have produced a programme will indicate that you expect some co-operation with meeting the dates. However, do remember that your programme is based on *anticipated* timescales – although you will have a certain amount of control over those you are employing, when it comes to planning committees you are likely to be in their hands.

SAMPLE PLANNING STAGES PROGRAMME (GUIDE)												
ITEM	Jan	Feb	Mar	Apr	May	Jun	Jul	Aug	Sep	Oct	Nov	Dec
Decide on requirements	■											
Arrange finance		■	■									
Source professionals			■									
Produce rough sketches				■	■							
Outline drawings for planning					■							
Apply for planning						■						
Prepare specification							■					
Material quantities & suppliers								■				
Contractors' details for quotes								■				
Receive planning permission									■			
Negotiate with contractor(s)									■			
Start project										■		

Main programme of works – long term

Producing an accurate programme is very important. It can take some time to put together, so a serious approach must be taken towards its preparation. When we look at the project, there will be particular considerations that will dictate the timescale of each element. The programme will have to be produced in conjunction with the physical ability to carry out the work, as well as the timing of material delivery. Some specific materials may have a long delivery period, which will dictate when certain elements can be undertaken. This is not an unusual situation; therefore it is very important to plan your project and the elements around the critical dates of material delivery.

When producing a programme, you must be realistic and take everything into account – including the weather. In the early stages of a project, a spell of bad weather could result in lost time that cannot be easily pulled back, if at all. On long-term projects you may be able to regain some lost time by increasing the workforce or working additional hours or days, but on short-term projects this is more difficult. A short-term project is more vulnerable to running over its expected budget if the duration is extended.

The programme will tell you at a glance how progress is being affected by delays, and what likely effect it will have on following trades. When you issue programmes to particular individuals who plan to arrive at a given time to carry out their work, they will not take kindly to being turned away because the area for their particular work is not ready. If this happens, you may find that it is difficult to get them back on-site when you need them.

> **It would be beneficial** to hold a meeting with the principal contractors to introduce them to one another, although you may find that many reputable contractors already know each other. This is always a bonus as they will generally liaise well with each other to achieve continuity. By providing the individual contractors with the contact details of other contractors they will have to interface with, you will have put in place a system for avoiding the 'blame game' – contractors blaming each other for delays.

The following example will give you an idea of a simple bar chart programme which can be used for projects of all sizes. This is based on the construction industry format for multimillion pound projects.

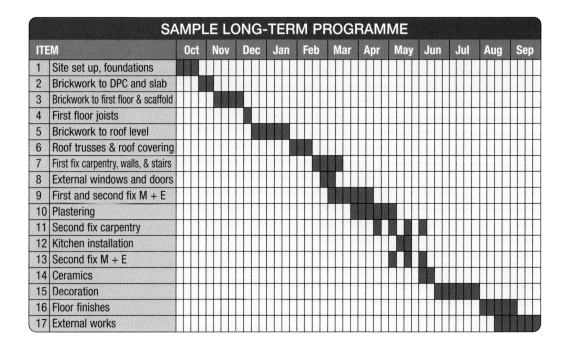

SAMPLE LONG-TERM PROGRAMME												
ITEM	Oct	Nov	Dec	Jan	Feb	Mar	Apr	May	Jun	Jul	Aug	Sep
1 Site set up, foundations												
2 Brickwork to DPC and slab												
3 Brickwork to first floor & scaffold												
4 First floor joists												
5 Brickwork to roof level												
6 Roof trusses & roof covering												
7 First fix carpentry, walls, & stairs												
8 External windows and doors												
9 First and second fix M + E												
10 Plastering												
11 Second fix carpentry												
12 Kitchen installation												
13 Second fix M + E												
14 Ceramics												
15 Decoration												
16 Floor finishes												
17 External works												

Short-term programme – for periods of high activity

Short-term programmes are extremely useful, as they provide a snapshot of particular timescales taken from the main programme. It is important to start using these from day one as they show clearly and precisely who will be on-site on any given day, and highlight any areas that may need to be discussed regarding continuity. Or they could alert you to the imminent delivery of heavy equipment, such as scaffolding or a crane, that has been organised by a specific contractor. This is vital to know as it will affect other work that is being carried out.

Another important use for a short-term programme is if the main programme has fallen behind. It will become your main tool for keeping track of progress, and show where you may be able to make up time by introducing a longer working day, weekend work, or by bringing in more workforce. As you can see from the example below, specific days for carrying out specific work can be identified.

SAMPLE SHORT-TERM PROGRAMME					
Week Commencing	23 Mar	30 Mar	06 Apr	13 Apr
1	Plaster living room				
2	Fix skirting to living room				
3	Fit fireplace to living room				
4	Plaster dining room				
5	Fix skirting to dining room				
6	Fit doors to living room & dining room				
7	Sample colours to living & dining room				

Material ordering and delivery dates

Once you have prepared the programme of works, you could produce a programme for ordering the material required. Bear in mind that if you know that some of your material is of a specialist nature or will need to be shipped in, they will need to be considered much earlier on in the project. By planning the ordering and delivery dates in detail you will avoid unnecessary waiting times and consequently the need to race around at the last minute picking up material. Most suppliers will run a delivery system. However, they will usually require a couple of days' notice depending on the type of material.

By contacting the suppliers to find out what is available 'off the shelf', and what will need to be ordered in, you will be able to provide continuity for each trade. Having a plan for material deliveries enables you to organise the unloading and storage requirements.

Payment schedules

Together with all its other uses, the bar chart is also used to plan an element for an action. By planning your cash flow, you will be able to monitor whether your expected costs are being realised. If the planned valuation dates start to extend, it may indicate

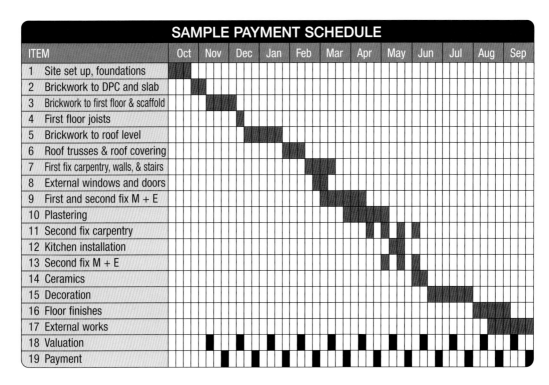

one of two things. Firstly, if the dates are not met it could be because the work has fallen behind due to bad weather, or target dates generally not being met. Secondly, it could indicate that additional work has had to be conducted, which could mean there will be extra expense. In either case you will need to give some thought to the long-term implications. If the project has suffered delays, you will need to consider how the time can be pulled back. If the delay is due to additional work being necessary, you will need to look at how the extra cost will be financed.

If your project is being financed by a mortgage company that has placed conditions on payments being made at set points in the project, the bar chart will assist in deciding when these payments may need to be drawn. This is illustrated in Chapter 1: *Sample Stage Payment Elements*. Alternatively, if you have decided to pay on monthly valuations, you can plot the payments on a schedule, which is simply a couple of lines on the bottom of the programme, indicating valuation and payment dates.

Training

Builders and contractors are legally obligated under the Health & Safety at Work Act 1974 to provide training for themselves and their employees for specific elements of construction. The type of training will be dependent upon the nature of work that they carry out, and the number of people that they employ. Unfortunately, training in the domestic side of the construction industry is not regulated to the extent that the governing bodies would like. This is a particular problem in the main building trades such as groundwork, bricklaying, roofing, and other non-regulated work. The only way to feel satisfied that you do employ qualified or competent builders or contractors is to ask for evidence of training, or to carry out good research before you commit to employing them.

If you are going to carry out work on your project as well as managing it, you must consider some training programmes for yourself in the use of various pieces of equipment. Whether purchased or hired, there are no specific laws on being trained in the use of some of the undoubtedly dangerous equipment that will need to be used. For example, there is nothing to stop you buying a chain saw from a tool supplier, but without specific training and instructions it could be lethal.

The site manager (or the person in charge at the work place) is generally responsible for the safety of the workforce under his control. There are specific obligations under The Health & Safety at Work Act 1974 regarding construction companies. (The domestic user does not have the same obligations and therefore things could become much more personal and complicated in the event of an accident, for example an individual may take legal action if they are injured.)

The person responsible for others on-site must ensure that tools are safe to use and that they are used correctly, whether hired or owned. He would inspect the tear-off stub on the hired equipment describing the last test/inspection and, provided it is valid, he can accept that the equipment is safe.

As far as the hire company is concerned, responsibility for misuse of their hired tools will fall directly to the person whose name is on the hire sheet.

Labour resources

Depending on your geographical location, you may find it difficult to locate suitable contractors to undertake certain elements of the project, particularly if you are new to the area. Now that computers are virtually a part of everyday life (and the Internet has expanded to a level where the smaller contractors can afford to advertise their services on-line), you will find a wealth of information, in addition to the methods that you would normally use such as Yellow Pages or other telephone directories.

There are on-line databases available for all manner of contractors you may need. In addition, local authorities are starting to provide a free service to their local communities whereby they provide information for specific contractors. They have a strict code of practice for people who want to advertise their services, but this is still in its early stages, and the local authority will not be held responsible for any subsequent problems that may occur. However, as time goes on the criteria will become stricter for contractors wishing to advertise on these sites, and this will reduce the risk of rogue contractors abusing the system.

There are other methods for finding reputable labour resources such as speaking to businesses in the local community. They may have contact with specific trades or contractors, estate agents or accountants. As most of the construction industry is serviced by self-employed personnel, accountants sometimes have a designated department or person who deals solely with contractors.

If you require general labour and do not have people you can call on, there are labour agencies that deal with providing the full range of labour that you would expect to use on a building site, such as:

- labourers
- ground workers
- drain layers
- bricklayers
- carpenters
- plasterers
- decorators
- foremen
- supervisors
- site managers.

There are other agencies that specialise in providing the professional services that are required in any building project, such as:

- project managers
- quantity surveyors – for calculating material
- estimators
- engineers – for setting out the work
- land surveyors.

As the labour element of all projects forms a large part of the cost, it is important to ensure that the right amount of labour has been allocated and the workload broken down in order for you to assign the work to the appropriate personnel.

Labour rates vary considerably across the country, and it is advisable to find out the appropriate rates for the area in which you are planning to undertake your project. You can find out the 'going rate' by asking at the local unemployment office, or by speaking to the various contractors who provide the services that you require. It is worth obtaining a few quotes from them. Some contractors with full order books may have a tendency to increase their rates slightly in order to take advantage of a boom period. This is a common practice and one that is unique to the construction industry, being one that suffers considerably from the effects of economic pressure.

Material suppliers

As stated in Chapter 1, the national suppliers will happily provide most of the material that you require for your project. However, they really do specialise in heavier building materials. By itemising the material requirements for your project into the different categories, you will be able to send copies to your local providers for obtaining quotes. If you prepare a list of material that you require, it will be easy for the suppliers to provide precise quotes, allowing you to compare costs. As you can see from the following example, the better you can put your list together, the easier it will be to make the comparisons. This could make a difference and produce unexpected savings.

	DESCRIPTION	SIZE (mm)	M²	LENGTH	TYPE	MAKE	QTY.	COST
	SAMPLE MATERIAL LIST							
1	Bricks	215 x 100	20		Eng			
2	Bricks	215 x 100	150		Facings			
3	Blocks	440 x 100	150		Concrete			
4	Blocks	440 x 150	60		Lightweight			
5	Joists	200 x 75		5.4m			40	
6	Wall ties			200mm	S/S	twist		
7	Studwork timber	100 x 50		3.0m			75	
8	Plasterboard	2.4 x 1.2	60		12.5mm			
9	Coving	100 x 120		3.0m				
10	Nails (round heads)	100					25kg	
11	Screws	100 x 10					400	
12	Screws	75 x 10					400	
13	Screws	50 x 8					200	
14	Skirting	150 x 25		4.2m	Chamfered		20	
15	Architrave	75 x 25		5.1	Chamfered		25	
16	Ceramic tiles	150 x 150	28					
17	Tile adhesive – non slip						50kg	
18	Paint (white matt)				Emulsion		75lt	
							TOTAL	

Use of technology

The management of a building project is now very much linked to the use of computer-generated material. This does not mean to say that your project will suffer if you do not use computers and computer-aided material. Everything that you see in this book can be produced by hand on graph paper and various formatted sheets that are available from good stationery outlets. However, one of the problems with producing material by hand is that it takes a long time, and if you need to update or amend something it can become quite tedious. For example, the bar chart programme is something that will inevitably need to be updated or altered as the project progresses and is more easily done using a computer.

Another problem with producing material by hand is that if your writing is not clear it may be misinterpreted. The printed word is obviously much clearer.

A further advantage of using the computer and associated software is that information can be sent electronically, speeding up the flow and therefore ensuring that orders are placed in good time. However, information sent electronically is not always viewed immediately, so it is worth following it up with a telephone call to ensure that it is.

You may feel that if you have to phone to check that your email has been received, you might as well dispense with the email, but the problem with relying solely on verbal communication is that it can be misinterpreted, forgotten or even denied. Sending a fax is a similar form of communication that minimises the risk of misinterpretation, but complications can often occur with fax machines. Although you may have a print-out of a date and time the fax was sent, it does not guarantee it was printed at the other end.

Most reputable material suppliers and contractors now prefer to use the Internet to communicate. There are computer applications that can assist in obtaining a receipt electronically. When you send a message, the recipient can send you a receipt or confirmation that they have received it by sending you a message that will automatically appear on their screen when they open the email. All they have to do is press the request for email confirmation in the middle of the screen.

The nature of the information being transmitted will dictate the necessity for backing it up with formal written and signed letters that may require a written response. Examples of this would be:

- contractual documentation
- drawings
- financial clarification
- certification of particular elements
- formal complaints
- changes to specification
- health & safety information.

There are many other examples which will become obvious by the exact nature of the project and the contractual arrangements. We will cover photographic records in Chapter 6. However, it is worth pointing out here that with the advancement of technology, digital cameras are now widely available at very reasonable prices. It is worth investing in one even if you do not own or use a computer. You can take hundreds of pictures of your project without having to have them processed. This may be vitally important in the event of any discrepancies. If an element of work has been obscured by the completion of other work and you have photographic evidence, this may be very useful for clarification. If there is a problem or situation that is of urgent concern, the pictures can be sent electronically. This gives the person at the receiving end a visual aid and therefore action may be taken more quickly.

Chapter 5

making it work

Plant hire and control

Plant hire can be a very inexpensive way of getting the work done without the need to spend huge sums of money on tools and machinery that may only get used occasionally. If you are carrying out some of the work yourself, it is likely that you will have your own small power tools. Small power tools are relatively cheap, and as they are used on a regular basis, it would not be economical to hire them over a long period. The types of tools that are worth owning are:

- circular saw
- percussion drill
- battery drill
- jigsaw
- chop saw.

Many of the expensive tools that are required for parts of a building project may only be used on a few occasions. These are the tools that are worth hiring, and each item is sent out with a ticket to show that it has been serviced since last being used. Also worth hiring would be items that are not easy to store, and therefore once used can be returned.

The types of plant and equipment that you would expect to have on-site from a hire company would be:

- Access equipment
  - Tower scaffolding
  - Ladders
  - Hydraulic working platforms

- Pneumatic breaker
- Electrical breaker
- Support props
- Water pumps
- Cement mixer

When hiring plant it is important to maintain control and ensure that only those who will be using the equipment concerned have access and permission to do so. In general, construction tools, equipment and material are not cheap, and can become a target for opportunistic thieves. For this reason you will need to plan for their safe storage if they cannot be removed from site. One of the safest methods of site security for material and equipment is a site container, which can also be hired. It is worth noting that if you do have a site container, small tools should still be locked in a tool chest, which needs to be secured.

Hire companies are responsible for ensuring the tools are safe for use, and inspections are carried out between hire periods. 'PAT' tests (a regulatory requirement for electrical safety and damage inspection) are conducted, and this is recorded on file. Hire companies have specific regulations to adhere to and are generally pro-active in advising customers of the safety factors. This may include written and/or verbal instructions, and (where necessary) demonstrations. For example, it would be normal practice for the hiring company not to hire a chainsaw with a guide bar length in excess of 355mm (14 inches) to anyone who could not produce a certificate to demonstrate proficiency in its use. For shorter chainsaws, the hiring company should exercise their judgement about the customer, and perhaps decline the hire for safety reasons.

A handover certificate would have to be signed by the hirer for large plant and for hazardous equipment. This certificate forces the hire desk to pass and receive more information and obtain signatures showing the customer understands and accepts these responsibilities. In order to minimise the risk of accidents, persons responsible for hiring any type of equipment should have good knowledge of his workforce's capabilities. He should also be aware of equipment exceptions, for example chainsaws: it is generally considered necessary for users of chainsaws to have undergone training.

If a member of the workforce expresses concern over his ability to use certain equipment, the person responsible should take appropriate steps either to train the person or give the task to someone with the necessary skills. This is a serious consideration as some of the tools that are used on construction sites are potentially very dangerous in the hands of inexperienced people. When it comes to large plant, the operator should have undergone specific training, which should be certificated. Although this may not be recent, provided he regularly uses similar plant and attends short update courses when necessary, you will generally be covering your considerations to health & safety in this respect.

Some heavy plant can be driven under a normal driving licence; however, this is borderline when it comes to driving it on a building site because of the obvious hazards.

Certain hazards, such as the dangers of machinery near excavations or overhead cables, may not be apparent to anyone who has not had any site safety-awareness training. The person responsible for hiring heavy plant must exercise his authority and allow only people he considers capable to use such equipment. Site rules should be introduced to ensure that no one uses equipment (including hire equipment) unless authorised to do so.

The following sample plant hire register will help to avoid the equipment being on hire for any longer than is necessary. It is easy to forget when an item was hired, particularly if you are hiring on a regular basis. By having the register on display and making a point of looking at it daily, you will soon get used to hiring and returning in accordance with your schedules.

DESCRIPTION		SAMPLE PLANT HIRE REGISTER				
ITEM		COMPANY	ORDER NO	DATE IN	DATE OUT	COMMENT
1	SECURITY FENCE PANELS	MARK 1 HIRE	22653			
2	DUMPY LEVEL	MARK 1 HIRE	23543			
3	OFFICE & FURNITURE	MARK 1 HIRE	24271			
4	STEPS	MARK 1 HIRE	25053			
5	ALUMINIUM TOWERS	MARK 1 HIRE	25068			
6	SKILL SAW 110V	MARK 1 HIRE	25085			
7	KANGO 110V	MARK 1 HIRE	25092			
8	HEATER	MARK 1 HIRE	25093			
9	DOUBLE EXT LADDERS	MARK 1 HIRE	25098			
10	DOUBLE EXT LADDERS	MARK 1 HIRE	25100			
11	GRINDER 225mm 110V	MARK 1 HIRE	25259			
12	TOOL VAULT	MARK 1 HIRE	25260			
13	ALUMINIUM TOWERS	MARK 1 HIRE	24777			
14	DRAIN TEST KIT	MARK 1 HIRE	25298			
15	GENERATOR	MARK 1 HIRE	45285			
16	BRICK-CUTTING SAW 110V	MARK 1 HIRE	25558			
17	ACROWS X 2	MARK 1 HIRE	25566			
18	TOWER SECTIONS	MARK 1 HIRE	25584			
19	WATER PUMP 110V	MARK 1 HIRE	25583			

This is an easy method of keeping a record of hired equipment, and it is important that all sections are completed. The section headed 'comment' is for noting if there are any obvious signs of damage, or if there are any components missing, for example from a tower scaffold.

Insurance for hired plant

For small companies or individuals, hire companies can offer simple insurance covering the equipment for the period of hire. Independent insurance agents may also offer schemes for this, but the hire company is worth considering as it does not make huge sums out of insurance and will probably only cover its administration costs.

Advice other than statutory H & S information

The hiring company will usually be happy to offer any advice and is encouraged to be pro-active in advising customers in any aspect of hire. This advice is based on training that they will have received and would normally be reliable and gained from experience.

If an accident does occur with hired equipment of any description, the hiring company has set procedures that they will follow to establish the circumstances of the accident in accordance with the health & safety regulations and their company's policy.

Site consideration checklist

In order to avoid being disappointed with the way in which site facilities are set up, it is important that you consider some of the risks associated with building work. Once you have studied the drawings and have established what will be required to run the project, it is worth having your own plan of how the site will be organised. It is not unreasonable to leave site planning considerations entirely to the builder's discretion, but by having your own plan of action that is open for discussion, the risks of being let down are minimised.

SITE CONSIDERATION CHECKLIST		
PARKING RESTRICTIONS	YES	NO PARKING ON-SITE, PARK CONSIDERATELY IN STREET
SKIP POSITION	YES	ON DRIVEWAY
PORTABLE WC	YES	ADJACENT TO GARAGE
MATERIAL STORAGE	YES	TO LEFT-HAND SIDE OF HOUSE
SCAFFOLDING – ALARM	YES	SCAFFOLDING TO BE ALARMED ON FIRST LIFT
SECURITY	YES	ALL OPENINGS MADE SECURE WITH LOCKABLE DOORS
FENCING	NO	EXISTING FENCING ADEQUATE
WATER FACILITIES	YES	TO RIGHT-HAND SIDE OF HOUSE
POWER FACILITIES	YES	TEMP BOARD TO BE FIXED ON OUTER WALL
NOISE RESTRICTIONS	YES	WORKING TIMES 8am–6pm, MON–SAT
DELIVERY TIMES & LORRIES	YES	BETWEEN 9am and 4pm
PARTY WALL INFORMATION	N/A	NO ACTION
TREE PROTECTION ORDER	N/A	NOT APPLICABLE BUT TAKE CARE NOT TO DAMAGE
LOCAL SCHOOLS	YES	CHILDREN PASSING PROPERTY 8am–9am and 3pm–4pm
CRANE REQUIRED	YES	BUILDER TO ADVISE ON DETAILS
DUSTY/NOISY WORK	YES	ALL DUSTY AND NOISY WORK TO BE CONTROLLED
PHOTOS	YES	PHOTOS HAVE BEEN TAKEN OF PAVEMENT AND ROAD
LOCAL AUTHORITY	YES	BUILDER TO INFORM LA FOR INSPECTIONS
RISK ASSESSMENTS	YES	RISK ASSESSMENTS REQUIRED FOR SPECIFIC WORK
METHOD STATEMENTS	YES	METHOD STATEMENTS REQUIRED FOR SPECIFIC WORK
FLOWER BED	YES	KEEP OFF FLOWER BEDS
PEDESTRIANS	YES	SAFETY SIGNS TO BE PLACED OUTSIDE
NEIGHBOURS	YES	CONSIDERATION FOR NEIGHBOURS' PROPERTY & CARS

Whether you are carrying out the work yourself or using a builder, if it is the first time you have had to think about the issues, preparing a site consideration checklist will help to identify the precautions that should be put in place and those that may need special attention.

When work is being carried out in an occupied property, there are specific health & safety precautions that need to be implemented. It is not unusual for clients to become distressed by the way in which the work is being carried out even if the standard is high. With refurbishment work or loft conversions where parts of the property will be exposed to the elements or particularly dusty conditions, you need to consider how particular risks of dust or other dangers will be controlled.

As you can see from the checklist, by writing down some of the issues that are important regarding how the project will affect yourself and others, you will soon be able to raise specific issues with builders, site personnel and/or contractors. It does not matter in what capacity people on the site are employed, there are issues on the checklist that apply to everyone. In order to understand the importance of such a simple but effective method of raising and controlling these issues, we will break each element down.

Parking restrictions

Unless you have a site that has ample room for parking, taking deliveries, and so on, it is an issue that requires thought and planning. The number of vehicles coming and going in even a small project can be substantial when you consider how many trades are involved. All tradesmen need to carry a number of tools with them and may even need to bring material on a daily basis. You need to decide who has priority for parking places based on whether or not a particular trade requires regular access to the vehicle. And it is important to think about how deliveries will be taken. If you do not have sufficient space for a lorry to unload (because your contractors are taking up too much space) this could cause traffic congestion and pose a health & safety risk. It is not always practical to keep a space for deliveries, particularly in built-up areas. Other people will naturally want to park their vehicles and by rights you cannot stop them. Placing signs and cones in areas where you need to park or unload can work; however, speaking to neighbours and people who will be affected will give you some idea of the amount of co-operation you can expect.

Skip position

Most projects will require some sort of waste-disposal system. The skip is by far the most practical method, particularly on domestic projects. Skips can be placed in the road provided that a licence has been obtained. The skip company will usually organise this; it is illegal to put a skip on the road without one. The cost of the licence is normally included in the cost of the skip itself. In general, skip lorry drivers will understand what is required of them and will abide by the regulations they have been trained to follow.

Some things to consider when placing a skip in the road

Other people sometimes use your skip for their rubbish: Bear in mind that when placing a skip on the road it is generally an open-top container and is therefore very easy for others to drop rubbish into. This can be minimised by placing the waste material to one side, and having the skip delivered when sufficient waste has been stock-piled. This method will work, though it means the material has to be handled twice, and so the cost and any risks must be considered. Placing a tarpaulin over the skip and tying it down with a couple of timbers placed across it (to prevent it sagging) will deter people from making use of your skip.

Skips on the road may need to have flashing lights on them: Depending on the local authority regulations, a skip may be required to have flashing lights fixed to it at night. You may find that in the small print of your contract with the skip hire company they have specifically pointed out that lighting is the responsibility of the hirer to provide and maintain. In this case, the skip hire company should be able to refer you to a supplier of these lights. Local authorities usually insist that as a minimum there are two reflective strips on either end of the skip and that each end of the skip is painted yellow. It is advisable to check with your local authority as to the regulations that apply in your area. Failure to comply could invalidate an insurance claim.

The skip company can insist on level loads: Although most people tend to make the most of a skip and load it over the 'level load' mark (usually printed on the skip), the skip companies generally use their discretion. Provided that the waste is secure and not too much higher than the skip edge, they will take it away. It is important that common sense is used when loading skips; the lorry driver may refuse to take it if the load is unreasonably high or unsafe. He may even unload some of the material himself and leave it on your property, so you could end up with other people's waste dumped on the site.

You may need a skip with a drop door if loading by wheelbarrow: Before ordering the skip, you need to make sure that the size and type will be adequate for the kind of waste that you are disposing of. For example, if all of the material is lightweight but bulky, you may need a high-sided skip with no door; but if you are disposing of earth from a footing, you may need a skip with a drop-down door to allow the use of a wheelbarrow to load it. Do remember that when you are calculating the size of skip you require, there will be an element of 'bulking up': voids within the waste material that could be as much as 35 per cent more than the measured area of original waste.

Your skip needs to be outside your own property: It would be very inconsiderate to place a skip outside another person's property, particularly where parking spaces are at a premium. Therefore you may need to plan ahead to ensure that a place has been reserved for your skip. This could mean that you need to put notes on cars and through your neighbours' letter boxes. It would also be advisable to have a vehicle parked in the place where you want your skip, in order to reserve the space. You could use cones to prevent people parking; however, unless the cones are official not everybody will co-operate.

Skip lorry drivers are generally very competent at offloading skips. They may have rollers that they can use to manoeuvre the skip into position. When it comes to loading the full skip on to the lorry, they usually prefer to 'back up' to the skip so that it can be lifted on in one movement. It is not possible to get rollers underneath it when it is full. They do have methods of manoeuvring full skips but it depends on the weight of the skip when it is loaded. It would be useful for the driver if there were at least two average car lengths clear at one end, in order for him to load it on to his lorry safely.

Skip on driveway: If you have a driveway on which the skip can be placed, do make sure that you are not causing yourself additional problems by taking up space for long periods of time, particularly if you need to get material or large items of hired plant in. Do not expect a skip lorry driver to place a skip on a driveway that does not have a purpose-built crossover between the pavement and the drive. The damage that can be caused to the pavement can result in the company being fined and/or charged for the reinstatement work, which can amount to thousands. If you are going to have on your driveway a skip that is going to be filled with earth, bear in mind that the weight will be excessive. When the hydraulic legs are lowered on the lorry to protect the lorry's suspension, there will be 'point loading'. This can cause damage to your driveway. The driver should place 'spreader boards' under the legs, which will distribute the weight sufficiently to avoid any damage.

INCIDENTAL CONSIDERATIONS

- Narrow roads will require more thought, particularly for emergency vehicles.
- If you live near or on the corner of a road, you may need to consider other methods of waste disposal.
- When loading the skips ensure that any vehicles parked close by are not damaged.
- You may want to cover vehicles to avoid any of the dust created when loading the skip.
- Always obtain the owner's permission before covering their car or give them the opportunity to move it.
- If a car does become dirty as a result of your skip being loaded, hosing it down (but not using a cloth) will remove the worst, and will help to avoid any scratching.
- Where there is a parking meter outside the property, you can suspend the bay. However, you may be charged a considerable amount for the loss of revenue to the local authority or controlling body.

Other methods of waste disposal

Waste disposal trucks

There are companies who provide a service to remove waste where a skip is not a viable option. They will arrive with a lorry, load up your waste there and then, and take it away. This may mean stock-piling but it is sometimes the best option. First, however, you need to satisfy yourself that they are legitimate, and hold the relevant licences. Your local authority may be able to provide you with a list of operators who are licensed to carry out this work.

Recycling

It is worth recycling some of your waste. Waste-management companies are encouraged to offer services and good rates for different types of material. If you call your local authority, you may find that there are schemes for recycling that specific companies participate in. Salvaged material is worth consideration, and examples of these are covered in Chapter 6. If you have large amounts of a specific type of waste (such as hardcore or topsoil) try telephoning several suppliers of this type of material to see if they will collect your waste free of charge. There are other types of material that may be passed on to a third party (such as redundant copper or lead pipes), which may even bring in some extra cash.

Portable WC

In the case of a refurbishment or extension, you may decide that the contractors can use your private WC. However, if you do have the facility to place a portable WC on-site, it is a good idea to insist upon it. If you can limit the number of people entering your property to those who need to gain access for specific reasons, you will eliminate fundamental problem areas. For example:

- dust and dirt being brought in to the house
- security problems
- misuse of your personal space.

If a portable WC is not possible, you need to ensure that there are strict rules regarding people entering your house to use your WC, to avoid people coming and going as they please. A procedure for access to your WC facilities could form part of the main site rules. However, this may seem excessive in the eyes of some builders or contractors and they may not at first take you seriously, particularly if they have not had this sort of regulation imposed upon them before.

Even if the property is going to be empty during the work, a portable WC is still worth considering – the most professional companies can have employees who will use a household WC, even if it does not flush! Since refurbishment work often involves sanitary ware and pipe work being replaced, removing the toilet completely will avoid this situation. In winter, removing the WC and its water feed will also minimise the risk of flooding due to damaged or frozen pipes. Portable WCs can be temporarily connected

Diagonal bracings should always be used for stabilising the scaffold

to a water feed from an outside tap using polypropylene pipes; these will not easily be damaged and can be lagged to avoid the water freezing. The WC can also be temporarily connected to the main sewer by connecting a 100mm diameter pipe to the rear of the unit and discharging it into an adjacent manhole. This eliminates any costs the hire company will charge for emptying it.

Scaffolding – health & safety, and security

The use of scaffolding needs to be discussed from two points of view: health & safety and security. Concerning health & safety, as discussed in Chapter 6, there are many issues that builders and contractors should be aware of and address, and the first consideration must be the risks involved in the erection of a scaffold. The second consideration is to find out what measures are in place to ensure that your security is maintained. Alarming a scaffold is expensive, but will give you peace of mind. There are other methods of making a scaffold secure, such as raising the ladder at the end of each day or fixing a board to the rungs to stop anyone climbing it, but an alarm is obviously the best option and thought should be given to this.

Material storage

Whether or not you are supplying the materials, the storage needs to be planned. Material that can be damaged by severe weather conditions should be covered or stored in a dry place. As the materials represent a major cost on the project, you need to allow an element of time each day to ensure that waterproof covers have been made available and are used. If your builder or contractor is making little effort to protect the materials, you need to raise this with him, as they are your property. You may meet with some resistance or be given a fairly plausible excuse that it will dry out, but there are factors that can cause problems once the material is fixed or when it is being fixed. Some examples of this follow.

Bricks and blocks

Rain on stacks of bricks and blocks that are not covered with waterproof sheets can cause the mortar to run when they are laid, which in turn can stain the brick face. The excess water in the brick or block can weaken the strength of the mortar. It can also have an effect on the colour of the mortar: when the wall is complete and has dried out, there can be obvious inconsistencies. Discolouration may not be apparent until after you have paid for the work, particularly if conditions are damp for long periods of time, making it difficult to get the bricklayer back to carry out remedial work.

Timber

If timber is left to the elements, in very wet conditions it will soak up water very quickly, which will cause it to swell. When it is fixed in place and starts to dry out, it will shrink back and possibly twist or bow, causing undulations to a floor or wall. If the timber is to be left natural when fixed, any previous water staining will obviously look unsightly. Where the timber is going to be painted, if it is not allowed to dry out fully before the paint is applied the paint may dry out very flat and discoloured. Timber left out in the sun can also be badly affected. It will twist considerably if it dries out too quickly. It may appear dry, but if it has been dipped or impregnated with wood preserver it could be very wet inside. If possible, timber needs to be stored in the shade, have battens between the layers to help ventilate it, and be allowed to dry out gradually.

Sand, cement and plaster

Moisture can contaminate sand and make it difficult to use. Loose sand needs to be covered as it would not take long to block drains if it is washed away by heavy rain. Cement is another material that often only gets scant attention paid to its protection. When cement becomes even slightly damp, it can become unusable. Plaster will absorb moisture very quickly and will need to be stored in dry, well-ventilated conditions. If you store plaster outside in damp conditions (even if it is covered) you run the risk of losing it all.

Plasterboard

Plasterboard needs to be stored under cover, at the very least in a garage, particularly in winter. It is very porous and will rapidly absorb moisture, causing it to lose its shape very quickly. When plasterboard is positioned against a wall, it needs to be supported by timbers to stop it from bowing inwards. Once it has lost its shape and then dries out, it will not easily flatten out when fixed and will crack if forced. If plasterboard is used when it is damp, it becomes difficult to use and is easily damaged.

Chipboard

There are many types of board material for flooring; chipboard is specifically made for floors and there are different types. Because of the way in which it is manufactured, it is fairly durable and does not easily absorb moisture. However, if it does get wet, it will swell and become unusable, as the water breaks down the formation of the material.

Paints

Liquids such as paint, varnish or glue need storing in dry and well-ventilated areas. Frost will soon spoil these and if tins, drums or containers are left in bright sunshine, they can heat up and explode. Another consideration is that paints can give off high levels of dangerous fumes which may be flammable or could affect your health.

Security of personal belongings on-site

Everybody involved in the project will readily tell you that you have no worries regarding security. However, problems do occur, and if you have no rules and have not taken measures to minimise the risks, you and everyone else will be very disappointed and the project will suffer from a severe lack of trust by all parties. You will need to organise rules for your own family if there are workers coming and going, who may need access to all areas in order to connect pipes or make routes for new services.

No matter what type of work you are undertaking, there are likely to be times when you are not there, and on some projects the property may be empty for long periods. Empty properties are a target for various groups of people who can cause misery, such as vandals who break in for no reason, or those who break in to steal tools or fittings. Proper attention needs to be paid to how openings will be secured, and perhaps a temporary alarm system installed, with clear warning signs for those who may decide taking a look inside without your permission.

Fencing

Since building material and equipment are expensive, it is worth considering hiring lightweight galvanised mesh fencing. This type of fencing will make it more difficult for unauthorised people to enter the site when no one is there. On longer projects, erecting a timber hoarding with lockable gates will obviously make entry even harder; this will also create a barrier that will make it difficult to see what is on-site.

Water facilities

If you are starting a new build project, you will need to plan well in advance for the water authority to provide a supply to the site. This will eventually become the main feed for the property. On a refurbishment or extension project you need to provide an external tap, which would not be difficult for a plumber to install. Providing a separate external supply could avoid the need for workers entering the property, which in turn will minimise the risk of:

- dust and debris being brought in to the property
- damage to sinks, basins and taps
- heat loss

- flooding
- contamination to clean areas
- security risks.

Using a hose from an internal tap is an option, but it does have its drawbacks, such as the hose becoming detached, causing internal flooding, and there may be a security risk where the hose enters the property.

Power facilities

On new build projects, it is likely that the power supply will not be connected until the building is virtually complete and all electrical work has been tested. The local electrical supply company may be able to provide a temporary supply for a long-term project, provided that the criteria for housing the live head and meter are met. Alternatively, you could hire a generator to provide power for tools and lighting as and when required. On a refurbishment or extension project, the domestic supply could be used. However, it would be worth consulting an electrician to discuss the provision of a dedicated fuse, to minimise the risk of the whole supply 'tripping out'. Again, small and quiet generators can be hired that would provide ample power for the types of power tools used on-site.

Noise restrictions

You need to establish from the start exactly what the working patterns will be, particularly with regard to noise levels. There are many factors that need to be considered, with the location and environment being the key elements. Common sense dictates that in a built-up area, neighbours and local businesses need to be taken into account, and if you are situated near to a school, hospital, or other such establishment, it would be worth arranging for noisy work to be carried out at times when it will be less intrusive. It is not always possible to accommodate other people; but in order to maintain good relations with your neighbours, it is a good idea to inform them of the work that you will be carrying out. People are generally more accommodating if they know what is happening and how long it is likely to last. There are some methods that can be put into place to reduce the effect of noise on others such as placing temporary boards between the source of the noise and the area that you would like to protect.

Delivery times and lorries

As with noise restrictions, it can be difficult to stick rigidly to set times for deliveries. However, if your project were situated in an area that is extremely busy at specific times of the day, it would make sense to avoid those times. It is also important that your suppliers do not deliver material to your project when you are not there. For example, skip providers in particular are renowned for delivering skips at unreasonable times in the morning, which can cause ill feeling between you and your neighbours.

You need to make your suppliers aware that if your project is in an area that would not easily accommodate a large delivery vehicle, they may need to make an advance visit to the site to ensure that their vehicle will have sufficient room for access. Materials such as large quantities of bricks or roofing material would normally be delivered on large vehicles, so you could see if your local building supplier would be prepared to break it down and deliver it in smaller loads. This would obviously involve some cost; however, if the order is large enough the supplier may be willing to arrange it. Alternatively, you could arrange to have a smaller lorry available. At the point where the supplier's vehicle becomes too large and access is not possible, the material could be transferred (possibly in several loads) and transported the final distance. If potential problems like these are not anticipated, days can be lost, as drivers will take the material back to their yard until other arrangements are made.

Party wall information

If you are carrying out any work to a party wall, you may need to consult a Party Wall Surveyor. They will ensure that the proper procedures are followed in order to comply with building regulations, and avoid or resolve any disputes. This is covered in Chapter 4.

Tree Preservation Order

Tree Preservation Orders (TPOs) are covered in detail in Chapter 4. However, even if you have trees that do not come under specific protection, it is very important to plan for any work likely to be carried out near them, whether it is above or below ground. Remember that if tree roots are removed or damaged, the stability of the tree could be affected, which in turn could present long-term problems and risks to your property.

Local schools

Even if you do not live close to a school, you may live in a street that is used by many children as a route to and from school. Children can be at particular risk as they may not easily recognise the dangers associated with building work: for example, when passing site entrances, they need to be made aware that there are vehicles frequently coming and going.

If a crane is required

If your project requires the use of a crane to unload material or lift a particular object into place, you will need to plan for this well in advance. If you are using a builder or specialist contractor, it is likely that they will have made all the necessary arrangements. However, if you are responsible for arranging a crane lift, you will need to speak to your local authority to find out exactly what the procedures are for obtaining appropriate licences. You may need to apply for a road closure, which will take time to arrange. There are many laws and regulations regarding crane lifts which need to be complied with. The selected crane operator would normally make a visit to the project to look at any potential problems that need addressing such as the need for a road closure, overhead power lines, narrow roads, etc. The crane operator may well organise all of the necessary legal formalities.

Dusty and noisy work

Of course, by its very nature building work involves an element of dust and noise. However, in order to minimise any health & safety risks, control measures need to be planned. The nature and duration of the project will to some degree dictate what areas need to be addressed. It is also important to consider how people who are not directly involved in the project will be protected.

Photographs

The importance of keeping photographic records is discussed in detail in Chapter 6. As a very minimum you should take photos of the condition of pavements and roads outside of your property, particularly if there is damage prior to your project starting. If you have contributed to causing damage to pavements or roads during your project, it could cost you a great deal for the remedial work, which can only be carried out by appointed contractors.

Local authority

There are many reasons for having contact with the local authority during any project that has to have planning permission or meet with building regulation approval. It is important to have elements checked by the Building Control Officer as and when required. It is therefore important to know who will be contacting the appropriate department or person who will carry out any checks.

Risk assessments

Building involves methods of work that need to be conducted with minimal risks to health & safety, or controlled by knowing the risks involved. Risk assessments will be explained in Chapter 7, but it is important to understand that where there are significant risks, there are specific ways in which they can be highlighted and understood.

Method Statements

It is important to know how a specific element of work that has been highlighted in the risk assessment will be carried out. Method Statements will also be explained in Chapter 7, and although you may not need to produce them for much of the work, it is important to understand when they will be required and the information that needs to be provided.

Flower beds

If you have areas that you do not want damaged (such as garden lawns or flower beds), it is important to point this out to all site personnel before materials or vehicles cause damage that can be expensive to rectify.

Pedestrians

Whatever type of project you are undertaking, the likelihood is that pedestrians will pass your site entrance. It is important to pay particular attention to the methods used to alert pedestrians to possible site traffic movement. Overseeing the movement of traffic on and off the site or the unloading of material across a footpath will minimise the risk of endangering passers-by.

Neighbours

Immediate neighbours may suffer from the general inconvenience of the work; discussing these issues with them will minimise the possibility of confrontation and upsetting them before you have even moved in.

The site consideration checklist does not necessarily need to be approached in too formal a manner; however, it would be appropriate to have early discussions with all concerned, before the project starts. As contractors start to arrive on-site to carry out their work they will need to know the site rules and any specific issues that apply to them. The checklist will be useful to them, in particular if specific elements related to their work are highlighted.

Keeping records

Record keeping for any project is where time and money matters are the key points to consider. All parts of the project will have a bearing on the projected forecast of your budget and the time that has been allocated for each one. If you start to lose control of the time factor for a specific element of the work, it can and will have a negative effect on your budget and the continuity of work. Things do go wrong at times which may not be attributable to a lack of planning, and should this happen it is important to know how to deal with it and whether you need to re-programme or plan to make up for any lost time. And good record keeping will minimise the risks of overspending.

There are many methods of recording details common to construction projects in the commercial sector, and these methods are now being introduced to the domestic sector. Not only is it important to record the necessary details, it is also important that the information can be easily retrieved. As you will be dealing with a variety of people and companies (from professionals to site labourers and national material suppliers to the little shop on the corner), it is important to have systems for handling the many different instructions and orders that are generated.

Some of the people and companies that you will be dealing with will have accurate systems for recording elements of the services that they provide. This will make your job easier when it comes to settling accounts; however, it is advisable to check everything against your instructions and orders, as mistakes can be and are made by what appears to be well-designed systems. For example, large suppliers will usually be prepared to negotiate good rates of discount with you if they are going to get a considerable amount of your business. However, human error is often the cause of these discounts not being passed on and you may receive invoices for material at the standard rates, so having records of your own to check is essential.

As you will see from the following methods of recording and passing on instructions, there is very little room for error. These systems are worth using for small as well as large projects. Contrary to popular belief, a small project does not require a correspondingly lower degree of record keeping, in fact due to the shorter working period, the project would need to be very tightly co-ordinated in order to maintain continuity. Co-ordination of any project can only be effective if the correct information is specified in the first place, and subsequent information, such as any changes or revisions, recorded in the proper manner.

Contractors who have priced to visit a project for a set period of time to carry out their work will not want to spend additional time if the cause of the extended time is due to poor co-ordination, for example, if they have been delayed by another element not having been completed, or by another contractor not fulfilling his obligations with regard to the programme. In this event, the contractor who has been delayed would expect to be paid for returning to complete his work. Situations like this will be avoided by making frequent reference to your records and making sure everything is going as planned.

As stated in previous chapters, it is important to provide sufficient information concerning timescales to all concerned prior to starting work on-site. Where a project does have elements that are critical and where specific dates need to be confirmed, it should be pointed out that confirmation of target dates will be given a week or so before the contractor is due on-site. The short-term programme (page 147) will help to identify these dates and will also assist in the dates being met.

Keeping a site diary

In order to keep a written account of your project (irrespective of the size or value), entering some of the basic information in a site diary will always ensure that you have information to refer to that will correspond with specific dates and times. The diary is a very useful tool and can be used in conjunction with some of the other systems that need to be put in place to control the project, such as ordering materials.

The best type of diary to use for any project is one which has an A4-size page to each day: these diaries tend to be well designed and may have month planners and sections in them for logging expenses, telephone numbers, emergency numbers, etc. Even if you are keeping records on a computer or laptop there are specific details that you should write down in your diary. Computers can sometimes fail, and it is not always practical to have a laptop on-site. There are some useful battery-operated digital voice recording devices on the market that are inexpensive. These are useful for making your notes while on-site, and you can transfer them into your diary at a more appropriate time.

Developing a routine will help relieve the stress of having to remember! By recording information whenever possible, you will minimise the risk of forgetting information that has a time or financial implication. Even if you are not in the habit of writing, without making this a part of your daily routine you will find it very difficult to become more efficient. Very few people have the ability to remember everything that they have to, what they have done, and all the people that they dealt with on a particular day. If you are working on your project as well as managing it, write down the morning's activities or notable events at lunchtime, or at the end of the day. Think of the importance of having a written account of, for example, significant events that may have led to the late completion of a certain part of the project.

The diary can be used for entering dates on which you need to be ordering material, or dates for when material is expected to arrive, allowing you to plan for additional labour to assist in the unloading and storage procedure. The weather should always be entered into the diary, as this is a factor that may be useful for identifying where time has been lost. This is very important as you may be issued with invoices from contractors who were held up due to inclement weather. It is important to clarify the position here as some contractors will not hesitate to claim for down time on a day-work basis.

You could produce a specific daily diary sheet that can be included in your site file. These sheets are useful on the larger project as they can save you time in writing down the same information daily. But it is still advisable to record information in your site diary. The following example will help to identify the key points to record on a daily basis.

SAMPLE FORMATTED DAILY DIARY					
DATE:					
ADDRESS:					
VISITORS:					
PURPOSE OF VISIT:					
INSPECTIONS:					
PHONE CALLS					
1					
2					
3					
4					
5					
6					
LABOUR ON-SITE	Nos	HOURS	CONTRACTORS	Nos	HOURS
GROUNDWORKERS					
CARPENTERS					
BRICKLAYERS					
PLASTERERS					
PLUMBERS					
LABOURERS					
ELECTRICIANS					
SCAFFOLDERS					
ORDERS PLACED:					
NOTABLE EVENTS:					
WEATHER: Rain Sun Dry Frost Windy					

As you can see from the example, there are some basic details that have been included. It is worth mentioning here the importance of including in your diary the time and purpose of any visitors, and in particular if the visit involved an inspection of some kind. If you have had a visit from the Building Control Officer who has checked out an element of work that is required to be inspected under the building regulations, it is a good idea to take some pictures of the work to back up the visit and make some notes of his comments.

You will make many telephone calls during the course of your project, the details of which may need to be recorded. For example, if you have agreed specific details with a contractor or placed an order for material, it is worth making a brief note of these calls, which you may need to refer to in the event of a discrepancy. Similarly, if you make a call to arrange a visit for an inspection to be carried out, recording it formally will prompt you into making a note in your site diary, which will help to ensure that your attendance at the visit is not forgotten.

Visitors

All visitors to site should be recorded in the diary particularly if inspections have been carried out. You may have engineers, architects or other professional consultants on-site who are meeting to discuss how to get over a particular problem or to re-design part of the project. It is therefore important to note who was there and when, what the nature of the visit was, and what conclusions were decided. It could be crucial to the information that follows, and the actions that you may need to take. If you do not have records of such visits you could find that delays in the flow of information start to occur. This is partly due to the fact that architects and engineers have other clients and may not respond to you in the timeframe that is needed to keep your project moving. When visitors are on-site and they have obligations to provide you with information, make sure you make notes that you can refer to for writing out your progress report or CVI (Confirmation of Verbal Instruction, see page 178).

Maintaining a site file

If you are employing a builder to carry out the majority of the work, you may only need to prepare one main working project file. Whatever size project you are undertaking, your working project file (more than one for larger projects) needs to be put together in sections, with each section describing the service being provided. As an example of this, the following file (which we will break down) will help you to formulate your specific type of file.

File 1: Main working project file
1. Contact details of all project contributors
2. Drawings and sketches (main works)
3. Engineer's drawings
4. Specifications
5. Programme of works

6. Correspondence to/from Architect
7. Correspondence to/from Quantity Surveyor
8. Correspondence to/from Engineer
9. Mechanical and electrical (M + E) drawings
10. Correspondence to/from local authority, planning, etc.
11. Contract details and orders for contractors
12. Correspondence to/from contractors
13. Site instructions
14. Health & safety information
15. Progress reports
16. Minutes of meetings

1 **Contact details of all project contributors:** By now you should know which contractors and suppliers you will be using for the project, and it is useful to keep all contact details on one or two sheets at the beginning of the file. The reason for this is that even before you start your project, you will have accumulated a great deal of information. You may have a full lever-arch file, or on a larger project you could have several files. Lever-arch files are probably the best method for storing the project information; however, if you have to leaf through each section to find contact details or telephone numbers on a regular basis, it will not take long for the file to become worn. If you have the luxury of a site office it would be better to display a copy of the contact details on a wall and use a filing cabinet for storing your correspondence.

SAMPLE TELEPHONE LIST					
OPERATION	COMPANY	NAME	TELEPHONE/FAX	EMAIL	MOBILE
Architect					
Builder					
Electrician					
Plumber					
Scaffolder					
Engineer					
Material suppliers					
Hire company					
Local authority					
Water authority					

As you can see from the sample, by having the telephone numbers to hand and the name of the person to contact, you will avoid wasting time leafing through files and paperwork each time you need to contact someone in relation to the project. On a large project it may be worth preparing separate sheets for each of the different elements, such as:

- ❂ builders, tradesmen and plant hire
- ❂ local authority and utilities

173

- professional contacts
- material suppliers (general)
- emergency contact numbers.

2 **Drawings and sketches (main works):** Since the drawings will need to be referred to on a regular basis, it is advisable to take very good care of them and ensure that you have a spare copy in your site file available at all times. Working drawings that are used for reference on a daily basis inevitably become creased and torn in time. And when they are taken on to site for setting out, they can get so wet and muddy that they become unusable; but there are special transparent covers available that will help to protect them in such conditions.

If you have a large project or a project that is very technical and requires lots of drawings and details, you may want to consider buying a device for holding the drawings on a rack, which eliminates the need for folding them. This method is much more practical, and apart from the fact that less damage is done to the drawings, less time is spent in leafing through all of the drawings kept in a file. If you do not have the facilities for keeping a drawing rack you could keep the drawings in a completely separate file, preferably a box file to avoid the necessity of punching holes in the drawings. This may sound trivial, but if you punch holes in drawings or detailed sketches, you run the risk of losing small areas of vital information such as dimensions and angles.

3 **Engineer's drawings:** In much the same way as the architect's drawings, the engineer's drawings may need to be referred to for setting out or for working to on a regular basis. Since the engineer's drawings generally form a more detailed part of specific areas, certain sections of the drawings may need to be photocopied and issued to

A low-level tower scaffold

tradesmen in order for them to work to particular design requirements. It would be useful to keep a copy of a specific engineer's drawing with the appropriate architect's drawing. You should also have a register that relates to the date of the drawing and drawing number.

4 Specifications: Under this section in the file it is a good idea to keep a copy of all specifications that have been issued. It may be worth keeping a list that can be revised in a similar way to your drawing register. If you do change the specification for any reason, you need to inform any contractors of those changes or exclusions as soon as possible.

5 Programme of works: Your main programme of work and any other short-term programmes need to be at hand and referred to on a regular basis. They will form an important part of your agenda for meetings with all persons involved in the project. Revised dates that affect any programmes need to be updated to see what effect this will have on the planned work, and any new dates issued accordingly. You should always keep your original programme as this is used as a guide to what you initially planned, and will help you to identify why and where any time has been lost.

6 Correspondence to/from Architect: In Chapter 1 we discussed the *Personal Project File*. It is useful to keep a general correspondence section or file, depending on the size of the project. By breaking these sections down in the main working file, you will be able to find relevant letters and correspondence at very short notice, which can be a frequent occurrence. When changes are made and confirmation of details needs to be relayed to specific individuals, you may need to refer quickly to previous letters in order to understand the full picture. For this reason correspondence needs to be strictly maintained in date order.

7 Correspondence to/from Quantity Surveyor: If you are using a quantity surveyor, there will not normally be a great deal of correspondence. However, as the other information is very important and relates to most aspects of the project, it will need to be available for reference purposes. This is particularly important where changes have been made as you will be able to work out from the initial calculations what adjustments may need to be made with regard to material and cost.

8 Correspondence to/from Engineer: Whether you have appointed an engineer yourself or the architect has appointed one, it is important that all correspondence is available. As far as his drawings are concerned, it may not be practical to keep copies of them with the correspondence. In fact you will have copies of them in the earlier section: *Engineer's drawings*. However, if there are specific details within a letter that clarify points on a drawing, it would be advisable to attach a copy of the correspondence to it.

9 Mechanical and electrical (M + E) drawings: Unlike the architect's and engineer's drawings, the mechanical and electrical drawings are not normally required to be passed

Checking existing M + E dimensions in relation to those on the drawing

by the local authority, although the work does have to meet building regulation requirements. M + E drawings are usually produced in conjunction with the architect's drawings, and there would normally be considerable correspondence between the architect and M + E designer to ensure that there are no conflicting areas regarding overall design, and other aspects such as performance and aesthetics.

10 Correspondence to/from local authority, planning, etc.: This documentation is important during any project. You may be asked to provide confirmation of planning permission or a number of other planning issues. The Building Control Officer may wish to see your correspondence although he may have copies of relevant information with him in relation to his visit. Specific visits from the BCO will be explained later in this chapter.

11 Contract details and orders for contractors: It is important to be able to access the terms of any contracts placed with individual contractors. There may be specific requirements that need to be confirmed to site workers who are not fully aware of the contract details. Such information may be, for instance, that the contractor or his employees are responsible for removing their waste or redundant material from site as it is generated.

12 Correspondence to/from contractors: As the project progresses, general correspondence with the contractor(s) will be generated. This could be anything from questions to copies of guarantees or instructions for manufactured goods, such as windows. Copies of the contractor's insurance documents also need to be kept in the file. Where there are separate contractors being employed on-site, the correspondence for each contractor needs to be kept in separate sections to locate it easily when required.

13 Site instructions: All site instructions need to be kept in date order and separate from the general correspondence, as they will be required when valuations are conducted.

14 Health & safety information: A copy of the contractor's health & safety policy (or statement of the measures that will be taken to protect the workers, visitors and others) needs to be kept on file and updated as appropriate.

15 Progress reports: Copies of all progress reports should be kept in the file, as they hold vital information on where any delays have occurred.

16 Minutes of meetings: Minutes of meetings need to be referred to as soon as they have been issued. There are usually resultant actions that need to be taken by specific project contributors who attended the meeting.

File 2: Material orders and correspondence file

If you are supplying the material for the project yourself, you will need to have a good system for filing the mountain of paperwork that is generated by the delivery of material. When material starts to arrive on-site there will be delivery tickets and, in many cases where materials have been ordered and paid for in advance, the guarantees may be with the material when it arrives. It is important to prepare a file for the paperwork; it may be required to check against invoices that are 'to follow'. Some suppliers may send only part of an order, in which case this should be shown on the delivery ticket. Where an order is short of the number of items stated on the delivery ticket, it is important to write on the driver's copy exactly what was delivered.

Below is an example of some of the headings that can be used for the range of different suppliers that you may be using. As detailed above, it may be worth preparing a separate contact sheet for each of them, and have this at the front as well as in the main part of the file.

- List of suppliers and contact details
- Programme of delivery dates
- General building material – concrete, bricks, sand, plaster, etc.
- Specialist material main works – roof tiles, timber, fixings, etc.
- Windows and doors
- Sanitary ware supplier
- Electrical fitting suppliers
- Specialist electrical suppliers – data, communication, remote systems, etc.
- Door furniture
- Wall finishes – tiles, wallpaper, paint, glass blocks, etc.
- External work suppliers – fencing, plants/trees, garden ornaments, etc.

Issuing verbal instructions by telephone and in person

Confirmations of Verbal Instruction (CVIs) are one of the most under-used yet most effective ways of recording direct instructions, whether you are giving them or receiving them, particularly within the domestic sector of construction. If you are employing the services of a builder or contractor, you will eventually be in a position whereby you are required to give verbal instructions or answers to specific questions. Some of these questions or instructions may have financial implications and in order to maintain continuity of work, there may not be sufficient time to ascertain the appropriate cost implications. If your contractor has a system to record the information and requests a signature to carry out the work, you will need to have some idea of how much the work will cost. You could have a pre-arranged hourly rate for additional work and if so it is important that the timescales for additional work and material used are accurately recorded.

Trust is a big issue when it comes to additional work that needs to be carried out by a contractor to maintain continuity. Some companies use the opportunity to claim for additional work to make higher profits, and it is therefore important to try to eliminate this by producing a comprehensive specification that minimises these opportunities. If a contractor does not have a good system of recording verbal instructions, you will need to issue written details. Remember that some instructions could be to remove items from the specification, or to lower the specification, therefore making a reduction in cost that will need to be deducted from the overall price.

If you are giving verbal instruction or confirming details to a question over the phone, it is important to make a note of the conversation and to confirm the details at your earliest opportunity to avoid misunderstandings that can occur with verbal instructions, which can quickly lead to the client/contractor relationship breaking down. The person to whom you are giving the verbal instruction is unlikely to write down what you have told them, and when.

If a situation arises whereby the verbal instruction involves a considerable amount of additional work and there are heavy financial implications, you will need to know the likely additional costs as soon as possible and before the work starts. This will give you time to look at your budget and either agree to the work or make other arrangements.

> **CVIs** (or any method of giving formal instructions) are important. They may form part of any legal procedures in the event of a dispute.

Some builders rely on the fact that because they are already on-site, they will automatically be asked to carry out additional work should the situation arise. The danger of agreeing to have additional work carried out without knowing the cost is that you could be presented with a bill for work that is far in excess of its actual cost. If a builder has to carry out additional work and needs to bring in additional labour or make special arrangements to collect material at short notice, there may be a slight difference in the normal rate, but this should be marginal.

SAMPLE CONFIRMATION OF VERBAL INSTRUCTION	
Issued by:	Issued to:
Date:	Instruction number:
Site address:	
Details of instruction:	
Sketches:	
Materials used:	
Hours spent:	
Signature of person issuing instruction:	

As you can see from the above example, there would be little room for misunderstanding provided that correct and sufficient verbal information was given. In some situations where verbal instructions are given, they are for record purposes only and have no financial implications, for example:

- colour of paint
- height of door furniture
- spacing of shelving
- setting out positions.

It is sometimes important to issue CVIs of this nature in order to ensure that your specific requirements are passed on to the person carrying out the work. Builders will sometimes use standard heights or use their own judgement where you may have your own preferences. CVIs are generally for use by the builder rather than the client. When issuing CVIs, always keep a copy in your site file.

Issuing site instructions

A Site Instruction (SI) is slightly different from the CVI in that it is used to give instructions *on-site* for changes to the specification or for additional work, other than verbal instructions. One of the most frustrating parts of having building work carried out is when the initial budgets for the work are exhausted, due to unforeseen circumstances, before the job is finished. This is where these site management systems come into their own, as they will help to keep track of expenditure and also indicate the final cost of the job. As previously stated, it is important to have a method of recording specific information that you have given, particularly for changes in specification or additional work. The builder is the first point of contact for issuing instructions; however, there are often circumstances where decisions are made on-site to make changes to the specification, or to add something. In this instance, and in particular if the work is already under way, instant instructions need to be given.

The main contract or agreement is in itself an instruction to carry out the work as specified on drawings or specification. However, there are nearly always extras or changes to the project in general that will have time or financial implications. Some of the changes may be elements of the work that the builder has sub-contracted out, and as such the sub-contractor may not want to make changes or carry out additional works without agreeing it with the builder. In the event that the builder is not contactable and the work needs to be altered or additional work carried out, issuing a Site Instruction stating the changes will help to maintain continuity and ensure that any cost savings are noted or extras are paid for. The benefit of this system is that you have a record of what additional work is being carried out rather than being invoiced at a later date for additional works for which you have no details.

Site Instructions are a useful and effective way of avoiding misunderstandings and disagreements on agreed works between yourself, the sub-contractors and the main contractor. Most sub-contractors will have an identified role to play in any project; however, as a client overseeing your own project it is your responsibility to maintain continuity. If an employee of a sub-contractor is not prepared to carry out work that he has not been instructed to by his employer, the SI will usually allay any fears that there may be about the cost not being covered.

SAMPLE SITE INSTRUCTION	
Issued by:	Issued to:
Date:	Instruction number:
Site address:	
Details of instruction:	
Area or work:	
Sketches:	
All work carried out in relation to this instruction will be subject to a re-measure and will be taken into account in forthcoming valuations	
Labour content: Hours:	Days:
Materials used:	
Signature of person issuing instruction:	

The Site Instruction is similar to the CVI and could be used as an alternative to it. However, there are differences in that the CVI is specifically used for verbal instructions, and the SI can either be issued on-site or sent to the builder or contractor.

Requesting information

When the project is under way there will be many occasions when either you or a contractor working for you will require answers to specific questions. If for example these questions cannot be answered verbally or if they involve technical information, a system for requesting the information in a formal manner is needed. Your builder may have a system for requesting information from you such as a 'request for information' sheet, sometimes known as an RFI. This is a pre-formatted sheet that may have a list of questions or may be a simple query that requires a response from you, the architect or the engineer. Although it is normally a builder who would issue an RFI to you, there may be times when you need answers from the builder or architect, and by issuing a formal sheet you may help to avoid information arriving too late.

As you can see from the sample RFI, there are sections that require specific dates to be completed. It is important to give these dates as they will give the person who is required to supply the information a timescale for response. Many site managers in the

SAMPLE REQUEST FOR INFORMATION	
Issued by:	Issued to:
Date:	RFI number:
Site address:	

Details of request:

Date information required by: --/--/--   Date information received: --/--/--

Details of request:

Date information required by: --/--/--   Date information received: --/--/--

Sketches:

Signature of person issuing instruction:

BUILDING CONTROL OFFICER AND INSPECTIONS

Builders and developers are required by law to obtain building control approval. This is an independent set of inspections usually carried out by an officer from the local authority who is qualified and experienced to ensure compliance with the building regulations. There are also Approved Inspectors, and both may offer advice before work is started.

The Building (Local Authority Charges) Regulations 1998 enable local authorities in England and Wales to charge for carrying out their statutory building control functions in relation to the building regulations. There are three functions that relate to the charges: for full planning applications, building notices and site inspections. The 1998 Charges Regulations came into force on 1st April 1999 and require local authorities to fix and publish their own individual charges according to a number of principles, in particular that they should be based on the cost of carrying out building control functions.

When certain elements of the work have been completed or they are at a stage where an inspection needs to be carried out by the Building Control Office (BCO), it is important that you have given the required notice to the BCO. Once you have made an appointment for the BCO to visit your project to inspect the work involved, you must wait until the officer has made his comments before proceeding. If the officer does not arrive at the allotted time, it is advisable to clear with the department of your local authority responsible for these inspections what you should do, particularly where you have open trenches that need to be inspected. If you have, say, concrete lorries booked or tradesmen waiting to carry on with their work you could minimise any potential problems by taking pictures of the work that had to be inspected and allow them to continue.

Insulation is one of the key elements in meeting building regulations

commercial sector of construction use these forms on a daily basis, particularly with clients or architects who have a reputation for not responding very quickly. If you have had drawings prepared by an architect and there are areas that are not clear, the RFI may assist in getting the architect to clarify the position in time for the information to be used.

When you start using these systems, you will find that the professional teams will hold you in much higher regard, as the responsibility for producing the information is put firmly back in their court. They are trained to expect to use these simple systems and will be only too pleased to work with you.

Inspecting the work and tolerances

As stated in previous chapters, it is important to set standards and implement monitoring and inspection procedures for each element of the work. Some of the work will be inspected by the Building Control Officer and some may be inspected by the engineer. You may decide to employ the services of an independent building consultant to inspect the work on your behalf prior to agreeing valuations or Practical Completion. There is, however, a British Standard (BS 5606) that deals specifically with tolerances for construction, and this should always be taken into account when making inspections.

Prescribed work as detailed by the engineer should always be inspected for conformity by an experienced building professional before being covered up

Buildings and structures are generally identified by the primary construction material, such as a brick-built building, a steel-frame building or a timber-frame building. It would be unusual for a building to be constructed from a single material, particularly a domestic building. For example, a brick-built building may have timber or concrete floors, which may be either cast 'in situ' or formed from pre-cast units. In most cases a timber-frame building will have a brick or block work outer leaf, which could be finished off with render or left as finished face-work. In all cases the different types of materials will respond differently at various stages of construction.

For example, in the short term the stiffness of the material may be much less than that in the long term. All material will respond differently to temperature and changes in moisture content in the air. In addition to this, the fabrication and construction tolerances may vary considerably depending on the material being used. Architects, designers and engineers are trained to understand the processes of construction and what allowances to make for the interface between the different materials.

There are specific terms that refer to tolerances and the characteristics that apply, for example:

- stiffness
- creep
- shrinkage
- thermal properties
- moisture changes
- effects of movement
- sealants
- deviations and tolerances.

Stiffness

In order to keep weight down and at the same time use less raw material there is a trend towards designing thinner concrete slabs, despite the spans on some designs being longer. It is therefore very important that assessments of the likely deflections of members (the way in which materials/members respond under load/weight) are accurate. In order to achieve the stiffness for the design, the designer or engineer is required to have detailed knowledge of the 'elastic modulus' of the various materials used. There are tables that designers and engineers refer to in order to ensure that where different types of material interface, the correct allowances have been made.

Creep

Creep refers to the process whereby some materials continue to deform or move under the application of a steady load. Creep occurs in most building materials, but the extent depends mostly on the following factors:

- amount of stress
- direction of grain
- density
- moisture content of the wood and relative humidity.

Shrinkage

Not all building materials suffer from shrinkage. However, where moisture is present in material there will be an element of this. Concrete can take several months or years to shrink due to the slow loss of excess water in the mix and may therefore show signs of shrinkage long after completion of the work. The effects of shrinkage will be most noticeable in thinner elements of construction, in particular screeds and rendering. Severe cracking may be produced depending on the speed of drying out, particularly where it is restrained by the other parts of the structure such as where rendering is applied to very dry block work.

Thermal Properties

Thermal properties of all building materials are taken into account and calculated into the design of a project to provide the specific 'values' required to meet building regulations. The thermal properties of material will have a bearing on the rate at which it expands and contracts. This will contribute to the acceptable tolerance levels that are defined under BS 5606.

> **Although timber** does not shrink in the same way as concrete and clay-based material, it will continue to expand and contract with changes in moisture content. This may produce, for example, cracking of joints and binding of doors and windows.

Moisture Changes

The rate at which construction material takes on or loses moisture will depend on the level of shrinkage or expansion. When material is delivered to site it can be in a very different state from that of the environment where it will be permanently fixed. Until the material has had time to stabilise it may well shrink or expand considerably, and the designers would normally make allowances for the provision of suitable expansion joints. As the relative humidity of the environment rises or falls, material will expand or contract accordingly, which will be determined by several factors, such as:

- moisture content (generally)
- direction of grain (in timber)

- density
- size.

Effects of movement

Where different materials are joined together, significant stresses can arise unless appropriate movement joints are introduced. Where 'differential movement' has not been allowed for at design stage, this may lead to overloading of individual elements. Obvious signs of 'differential movement' include:

- cracking of brittle material
- curling up or warping
- binding.

Sealants

Sealants of all types have been formulated for use in various capacities and will usually be compatible with a range of materials that is specified on the container. The properties of the individual sealants will normally be significantly different from those of the elements or materials on either side, such as sanitary ware and ceramic tiling. The sealant's properties will have different characteristics such as strain capacity and stiffness, and will be chosen on the basis of the anticipated movements and environmental conditions that the sealant will have to accommodate.

> **Sealants must be used correctly.** Incorrect specification of a sealant (for example, one that is too stiff) may lead to damage to the elements on either side; or one that is designed for internal use may not provide appropriate weather proofing if used externally.

Deviations and tolerances

The 'perfect' unit or structure does not exist and deviation will include:

- variations in dimensions
- twisting
- out of upright
- out of level.

Errors can and do occur on a regular basis, particularly in the setting out stage of a construction project. However, in most cases these errors are minimal and are seen as acceptable tolerances. It is important to understand that variations in dimensions will occur at nearly all stages of the construction process, but as the project nears completion, the elements (building materials) will become smaller. Therefore the tolerances become smaller and will in many cases be unnoticeable.

Some timber joints come up slightly bigger than others, but can still be installed as the tolerances will be reduced when the finishes, such as plasterboard and plaster, are applied

The British Standards and approved documents provide the maximum permitted deviations for individual elements. For example:

- the National Structural Concrete Specification
- the National Structural Steelwork Specification
- BS EN 336, Round and sawn timber – Permitted deviations and preferred sizes, Part 1: Softwood sawn timber, Part 2: Hardwood sawn timber.

Different industries work to different 'normal' tolerances and therefore the accuracy of a construction project needs to take this into account. BS 5606, *Guide to accuracy in building*, gives guidance on manufacturing tolerances, for both in-situ construction and also for prefabricated components. The tolerances of off-site manufactured components will generally be better than those achieved on-site because of the use of standard jigs for steel or timber frames, moulds for pre-cast clay and concrete units, etc. Where brick and block work are concerned, coursing heights are maintained by adjusting the mortar joint thickness.

Movement joints are designed in many structures to allow for expansion and/ or contraction due to shrinkage, moisture and temperature changes. The timing and

Even if the work looks almost finished, payment should be held back
until full completion and inspection has taken place

application of finishes such as paint and ceramic tiles will depend on the substrate and how long it takes to dry out fully. This would normally be allowed for in the construction programme. One area to be aware of is the application of floor coverings to concrete floor slabs. The concrete must be allowed to dry sufficiently before the adhesive is applied, which can take several weeks or months. Even moisture-tolerant adhesive can be affected by moisture trapped in the floor slab, which can cause long-term problems with the floor covering.

It is advisable to introduce a quality check system in order to monitor the quality of work. This will ensure that each element has been completed to the standards that you have set, whether it is work that you are carrying out yourself, or work that is being carried out by a contractor. This is an important consideration as all too often the finishing of work is an area that can cause relations between you and your contractor to suffer. Similarly, if you are carrying out the work yourself, it is easy to get to a point where you have completed 95 per cent of the work, and then tell yourself that you can return to the finishing at a later date, only to find that other elements take over.

Contractors will quite happily present you with an invoice for work that is almost complete with the promise that they will finish the remaining work by the time they have received their cheque. Unfortunately, this too is an area that causes problems, as contractors may have set up their next contracts and give priority to them instead of completing the work on your project. For this reason it is important not to pay all of the money owed (less any retention) to a contractor if he has not fully completed his work.

> **If a contractor** has completed 95 per cent of his work, it would not be unreasonable to pay 75 per cent of the value of his work rather than the 95 per cent completed. The contractor will have more incentive to complete his work if he is owed what may amount to the profit element of his contract with you. When you use this system for each element of the work, you will find that most tradesmen and contractors will automatically co-operate with you.

Chapter 6

keeping costs down and staying safe

Structural changes and cost implications

Engineer's drawings will usually need to be verified and passed by the building control department responsible for checking the calculations against proposed architectural designs. Once all drawings have been passed and work is under way it is not uncommon to have to make minor changes to the engineer's drawings. Where this happens the revised drawings will be sent to the Building Control Engineer for approval. If the work continues before approval has been granted, it is done so at the risk of the client. As a general rule, small alterations to engineer's drawings would not necessarily warrant the project stopping unless confirmed in writing by the Building Control Office, although in theory you should wait.

Where major structural changes are required that differ considerably from those produced in the initial stages, it would be unwise to proceed with that element of work until the calculations have been agreed by the building control department responsible for checking the calculations. If you did proceed and the calculations were subsequently rejected, any work that had been carried out may have to be totally removed. Apart from the additional cost of physical changes to structural work, you may be faced with a substantial bill from the engineer for re-designing the work. This is where your

Original architect's plan showing structural work required

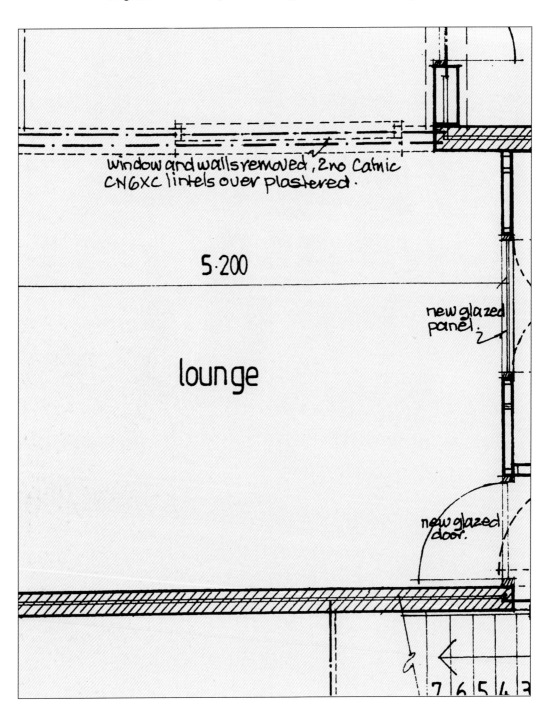

window and walls removed, 2no Catnic
CN6XC lintels over plastered.

5·200

lounge

new glazed
panel.

new glazed
door.

7 6 5 4 3

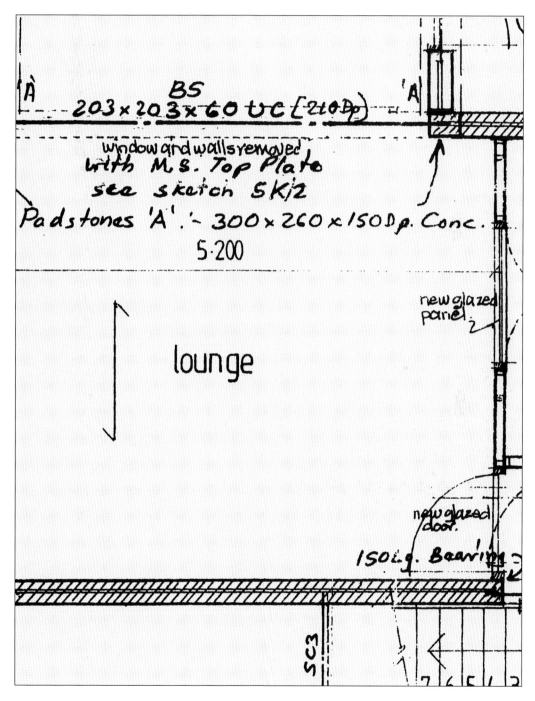

It is essential to ensure that any structual engineer's alterations are passed
by the building control department and updated on working drawings

The plans must be inspected to ensure that they match the existing structure.
If there is any doubt, the structural engineer must be consulted

contract or agreement with the engineer or architect needs to be fully understood. You may not have any choice in the matter of paying for redesign work; however, you may be able to negotiate a preferential rate for the work by anticipating the situation.

Issuing revised drawings

All drawings and sketches should be accompanied by a drawing schedule. It needs to be updated each time a drawing is revised, or if new drawings are introduced. When you issue revised drawings or copies of particular sections to site personnel it is worth making a note of it in your site diary, and also to make a note on the copied part in order to identify where the drawing it relates to came from. You will also need to retrieve the original drawings and mark them as superseded. You can keep these drawings for record purposes, as they can be useful for marking up ideas, etc.

When you receive your drawings they should be numbered and dated with a drawing register that shows which drawings are current and who is included in the distribution list. This register will need to be updated each time the drawing is revised. Many architects will automatically provide a drawing register and update it each time a drawing is revised, but if they do not you will need to discuss this with them. It can be difficult to keep track of drawings on some projects where details are changing on a regular basis. If you do not provide revised and updated drawings with the appropriate revised specification, you could find yourself with huge additional costs for abortive work. Contractors who are on an agreed price for their work will need to be informed of any changes to the specification in good time, in order to avoid unnecessary work and therefore additional costs.

Distribution of drawings can be fully co-ordinated by the architect, and in some cases, where other consultants' input is required, this will happen automatically. However, if you are using individual contractors, this may become costly and it will therefore fall upon you to ensure that information is passed on accordingly. If you do issue copies of drawings or sketches to contractors or site personnel, it is important to make a note of this in your diary, which should include the drawing number and revision number if appropriate.

You could make the builder or each contractor responsible for the cost of re-issued drawings, which would give them the incentive to take better care of them.

Holding site/progress meetings

On a small domestic project, it would be reasonable to assume that meetings would be held at the property in a fairly informal manner. The details of the work will obviously need to be discussed and recorded, but it is the management and legal obligations that also have to be discussed. Providing that the details of conversations and agreements are recorded, it is as good as a site meeting. It is important that the actions of the meeting are carried out and at subsequent meetings the progress of the project and other specific issues are discussed. If there have been changes to an agreement made at a previous meeting, this needs to be updated in order to agree the final cost implications.

When we start to look at projects of significant values there needs to be a more structured and planned approach to holding site meetings. It is important that the appropriate people are present at meetings where they are required to provide some sort of input, or for example to explain reasons for delays. Site meetings on reasonable-size projects need to be planned in advance, with the people who are expected to attend being given sufficient notice to plan ahead.

Principal contractors who interface with each other should be present to ensure that continuity of work is not affected by lack of communication. In their opinion, there may not be any apparent problems or issues that need to be resolved; however, there may be other issues raised at the meeting that they need to be aware of. The following are headings under which a typical meeting could be held:

- Progress against programme
- Elements of the project

- Health & safety issues
- External issues
- Delays
- Information required

Progress against programme

By producing a progress sheet for each meeting with the approximate percentages of work carried out, you will be able to see whether the project is slipping behind. You may need to request the percentages of work carried out by specific trades in order to show a more accurate picture. If you are employing a builder to undertake the work, you should expect to see a progress report and if you do not receive a 'progress against programme' sheet, it is worth requesting one, as visual aids such as these help to identify areas that may be causing delays.

SAMPLE PROGRESS AGAINST PROGRAMME SHEET				
Date:		COMMENT	Programme %	Actual %
1	Site set up, foundations	Complete, inspected and passed by BCO	100	100
2	Brickwork to DPC and slab	Complete	100	100
3	Brickwork to first floor & scaffold	Complete	100	100
4	First floor joists	Complete	100	100
5	Brickwork to roof level	Complete	100	100
6	Roof trusses & roof covering	Behind, due to inclement weather	100	95
7	First fix carpentry, walls, and stairs	Ongoing, no unforeseen problems	25	10
8	External windows and doors	Ordered, due on-site next week	10	0
9	First and second fix M + E	Ongoing	20	5
10	Plastering	Not started	0	0
11	Second fix carpentry	Not started	0	0
12	Kitchen installation	Not started	0	0
13	Second fix M + E	Not started	0	0
14	Ceramics	Not started	0	0
15	Decoration	Not started	0	0
16	Floor finishes	Not started	0	0
17	External works	Not started	0	0

Elements of the project

Elements of the project that need to be discussed will be dictated by the 'progress against programme' sheet, illustrated in the sample: *1 – Site set up, Foundations – 17 External works*. The important point to remember here is that if the project is behind schedule, the reasons why should be discussed, along with ways in which it can be brought back on to programme. It is worth taking into account that where you have made changes to the programme, or have instructed additional work to be carried out, pulling back time may not be easy to achieve. If an area of a project has come to a halt for some reason, this will of course have an effect on the completion date, in particular on smaller projects. If contractors have been programmed in to carry out specific work, they will need to be informed if the dates have changed. Larger projects may have the benefit of allowing work to continue in other areas. When formal meetings are held, it is usual for the contract administrator or person overseeing the project to take minutes of the meeting.

Health & safety issues

Health & safety issues need to be discussed. We think of health & safety as an on-site issue, but there may be certain points that you need to bring to the attention of contractors. For example, non-site personnel entering the site, or if there are particular activities planned that will create excessive noise or dust. If there are health & safety issues that you are unhappy about, you should initially bring them to the attention of the person responsible, then raise them at the meeting in order for them to be addressed and formally documented.

External issues

External issues cover circumstances that may need the co-operation of site personnel in general. For example, warning neighbours of activities such as crane operations that require full use of parking spaces, or trimming back trees, or it may be that site personnel are required to keep an area clear for a neighbour to carry out a specific task.

Information required

Although information would be requested during the project as and when it is required, answers may not always be provided when they are needed. The site meeting is the perfect opportunity for all parties to have any questions answered. It would be useful for the trades that interface on a regular basis to use the meeting to iron out any problems they may have.

Progress reports

Builders on fairly small domestic projects would not normally be expected to produce progress reports, and will not normally offer them as a matter of course; however, as the client, it would not be unreasonable for you to request a weekly or fortnightly report. The report need not be computer aided or in a very formal format: even a hand-written account of the summary of work shows that the contractor is aiming to complete the work by a given date.

As previously indicated, on larger projects, site/progress meetings will ensure regular dialogue and evidence of contractual documentation will assist in the flow of information being maintained. It is important that you as the client take full control of these meetings and have your correspondence and updated information at hand to back up any new instructions or changes in specification. As far as the builder/contractor is concerned, it is his responsibility to provide you with a full report on where the project is in relation to the programme.

This view demonstrates that the engineer's requirements for supports have been met

Photographic records

When you are planning to carry out any type of work to your property, or even build a new house, you need to begin with a mental outline of what the finished product will look like. Magazines and books will provide you with plenty of inspiration, and before too long you will have a pretty good idea of how to proceed.

Once you start your own project, it is advisable from two points to take plenty of pictures. Firstly, when you take pictures of the project from start to finish, it will give you a great deal of satisfaction to be able to look back on what you have achieved. Secondly, there may be times when you need to take pictures to produce evidence of work that hasn't been carried out to your specifications, or to record work that has been carried out in

It is always useful to keep a photographic record of the construction phase of your building project

accordance with the building regulations and that will later become obscured. With the age of technology now making it affordable and very easy to produce pictures at home, it is possible to take thousands of digital pictures.

Practical completion

Practical completion can be a very contentious term. I will give some dictionary definitions of the words *practical* and *completion*:

- Practical:
  - adapted or suitable for a particular use
  - relating to actual performance rather than to theory
  - very nearly as described or as described in practice or effect
  - a practical plan or suggestion is one that is useful, realistic and effective.
- Completion:
  - to bring to an end
  - to bring to a perfected state
  - to mark the end of.

The word *completion* cannot be argued over; however, *practical* is a little more ambiguous.

In the specification/scope of works and other contract documents, you may often see the term 'Practical Completion', a term that is often open to interpretation. The issue of whether a construction project has achieved practical completion has long been the source of many disputes within the construction industry, both in the commercial sector and, in particular, the domestic sector. This is not surprising, as many of the standard forms of Contract that are used (predominantly by contract administrators on behalf of a client), rely upon the issue of a *Certificate of Practical Completion* to trigger such matters as the release of the first half of any retention funds held.

Given its importance, it is surprising that the standard forms used throughout the construction industry do not define practical completion. They generally leave the matter to the discretion of the contract administrator, architect, engineer, supervising officer, etc. The preparation of a comprehensive specification/scope of works is paramount to minimising the risks of misinterpretation, although the risks will never be fully removed.

Various attempts have been made by the courts to form a definitive opinion for practical completion; however, it is likely that court cases will continue to be brought by dissatisfied clients who may have justifiable reasons for doing so. Practical completion of any part of the work in my opinion should be seen as the work having been completed to a 'set' standard where the client can take possession of the works and use them as intended. It is important for both parties (client and contractor) to use common sense when agreeing on practical completion. On the one hand the contractor should make every effort to complete the work to a satisfactory standard, and on the other the client should exercise some discretion on the finished product. That is not to say that poor standards should be accepted, but that some materials, such as bricks, timber, etc., can vary in colour or dimensions. This is where the monitoring and inspection procedures will help to ensure that the standards that have been agreed upon are achieved.

Judge Newy [in Emson Eastern v EME Developments (1991)] stated that:

> '...because a building can seldom if ever be built as precisely as required by drawings and specification, the contract, realistically, refers to 'practical completion' and not 'completion', but they mean the same. If, contrary to my view, completion is something which occurs only after all defects, shrinkages and other faults have been remedied and a certificate to that effect has been given, it would make the liquidated damages provision unworkable.'

Where a project reaches the intended programme dates and is offered to the client as being complete, it would be reasonable to accept the project as having reached practical completion even if minor works were outstanding.

Where explicit reference has been made to extremely high levels of finishes being required over and above what would normally be classed as acceptable, and whereby samples of work have been provided, agreed upon and documented, it would be difficult for an independent arbitrator to argue in favour of the client if the levels were achieved, and vice versa.

While practical completion seems to be generally understood by the industry, in practice there remains a difficulty in arriving at a formal definition that is accepted within the industry as a whole. It is generally accepted that:

- the Works can be practically complete notwithstanding that there are latent defects (present but not obvious)
- a certificate of practical completion or written agreement that the contractor has achieved practical completion may not be issued if there are patent (obvious) defects
- the Defects Liability Period is provided in order to enable defects not apparent at the date of practical completion to be remedied
- practical completion means the completion of all the construction work that has to be done as stated on the drawings and specification/scope of works.

Where a contract administrator has been employed and the project is being overseen under a formal contract, the administrator is given discretion under specific clauses to certify practical completion where there are very minor items of work left incomplete, on 'de minimis' principles. (The principle of 'de minimis' derives from the legal term 'de minimis non curat lex', and has been stated as meaning: 'the law does not concern itself with trifles'.)

It is surprising that the more frequently used standard forms of contract contain no clearer definition for practical completion, given the ambiguity in current case law. In the absence of such clear definition or persuasive case law, the main contractor is at the mercy of the client or contract administrator and their interpretation of practical completion, against whom he may have to argue his position accordingly.

Labour cost

If you are going to use labour-only contractors for some or all of the work, there are several methods of paying for the work, for example:

- price-work
- day-work
- hourly rate.

The labour content for building a new property is fairly easy for a quantity surveyor to calculate. Therefore it is more practical to have this work done for an agreed price. With other types of work (such as refurbishment and alterations) it is not so easy to pinpoint costs. By providing a good specification, it is possible to minimise costs. However, a 'labour-only' contractor may include a figure to cover unforeseen problems. If no unforeseen problems arise, the contractor will obviously profit. This is fair given that any problems encountered would have to be paid for out of his price.

Day-work is a good method of having work carried out, providing that the output of work is sufficient. If you are going to employ labour on a daily or hourly rate, it would be advisable to monitor the work closely, and if necessary employ a foreman to ensure that the output of work is acceptable.

There are national rates for all types of building work, and some of these are listed in 'The Housebuilder's Bible' by Mark Brinkley (6th edition, Ovolo Publishing, 2004).

Working hours

If you are planning to employ labour on a daily or hourly rate, you will need to decide on the hours of work for your project. As building work is physically demanding, working long hours can start to become counter-productive. The standard hours for building work are: 8 a.m.–5 p.m. (with a 15–20-minute break mid-morning and mid-afternoon), and a 30-minute break at lunchtime. Whatever hours you decide upon, it is important to maintain them, together with the duration of the breaks; when breaks over-extend, there is potential to affect the programme of work and also the budget.

Builders working on an agreed price are likely to have prescribed working periods which you may not necessarily agree with. However, if the project starts to slip behind for no apparent reason and there is the potential for more hours to be worked, you should raise this with the builder.

Material

It is very important to approach any project that involves spending your savings or borrowed money in a business-like way. Mistakes can become very costly when undertaking building work and they do not have to be big and obvious mistakes. Even small mistakes can soon start to eat into your budget without you realising it. There are certain areas that need to be monitored and controls put in place in order to avoid waste. An example of this is ensuring that your material orders have been properly planned, and that when material arrives on-site its distribution has been planned.

'Labour-only' contractors sometimes tend to use material that has been ordered for a specific purpose in totally the wrong areas. They may not be aware of where specific material is intended, and could quite innocently use the wrong material for the wrong job. For example, when timber is ordered, it is usual to order a slightly longer length than is actually required. This will allow for a small 'off-cut' to square off the ends. Examples of the types of problems encountered are:

- If you have ordered lengths of timber for an open cut roof (which may have varying lengths that differ by as little as 300mm) and they are used in the wrong places, you could end up having to buy replacement timber to achieve the required overlaps.
- Floor joists – it is not unusual for floor spans to be very similar from room to room, and if the wrong floor joists are used because they have not been correctly labelled, you could end up having to order new timber.

In these situations it is possible that a supplier would re-stock the unused timber. However, they would probably charge for this, and your replacement timber may not be in stock causing a delay to the work.

Ordering the correct quantities of material is very important. If you are not familiar with the terminology and sizes/lengths that are required, you may want to consider using a quantity surveyor to carry out a full inventory for all the material as specified on the drawing. Some suppliers offer a service where for a small fee they will work out the quantities of material that you require.

> **A builder** who is supplying the material and labour will organise everything in the package, but when you use individual 'labour-only' contractors, they will expect the material to be there when they need it.

Salvaged materials

Whatever type of project you are preparing to undertake, it is always worth taking the time to consider if there are any materials that can be salvaged for re-use. As there are many types of projects, there will also be many uses for material that you may initially believe would not be of any use to you. This is where you need to consider the project as a whole and see what you have to buy in for the various stages and what you can salvage for re-use elsewhere.

These projects could potentially involve using salvaged material:

- Building an extension
- Loft conversion
- Building a new house
- Refurbishment
- External works
- Demolition

Building an extension

If you are building an extension, it is likely that there will be the need for a new opening to be formed to open up an existing room on the ground floor, and at the very least a new doorway to be formed on the first floor. Typical materials that can be salvaged in this example include:

- **Existing bricks and blocks, which can be used for:**
  - hardcore in the base of an over-site
  - bricking up other doorways or openings, depending on how the bricks and blocks clean up
  - a base to form a driveway
  - fill to form the lower part of a raised flower bed (good for drainage).

  (If the effort involved in breaking up the hardcore is too great, you may be able to dispose of the hardcore free of charge if it is clean, i.e. free from other material such as timber, steel, glass, etc.) Suppliers of aggregate such as crushed concrete, which is used as sub-base material for driveways and concrete floors, may provide a specialised vehicle and driver to remove the hardcore, provided that the driver can get the vehicle close enough to use the hydraulically operated grab bucket.
- **Windows and doors:** Where existing windows and doors need to be moved, it is possible for a carpenter or specialist window installer to cut them out and re-fix them elsewhere, particularly useful where you are having windows made to match existing ones. This will need consideration when designing your extension but, depending on the size and condition of the windows or patio doors, it could represent a considerable saving.
- **Skirting and architraves:** Where a new opening is being formed, it is likely that you will need to remove sections of skirting or architrave. These salvaged lengths may be useful for patching in areas where doorways have been bricked up. The age and type of property will dictate whether the existing skirting/

architrave can be matched by 'off the shelf' material. In most cases there will be material that is the same or very similar, though new skirting sometimes comes up slightly smaller.

- ☻ **Flooring:** Where floorboards need to be removed, it is worthwhile punching the nail heads through the board, and salvaging the existing boards, as this can be an expensive element to replace. New flooring usually comes up narrower in width and shallower in depth, causing problems when laying floor coverings. If you decide to sand your floors and have not salvaged the existing floorboards, you can have new boards milled/planed to suit the existing ones, but again this can be expensive.

Loft conversion

During the building of a loft conversion, it is necessary to remove specific elements, for example:

- ☻ roof tiles
- ☻ roof timbers (rafters)
- ☻ ceiling joists.

Roof tiles are worth salvaging particularly if the plan is to match up the new work with the existing ones. If your loft conversion includes building dormers at the front and rear of the property, you may be able to use the salvaged tiles for one elevation, usually the front, which would help the new work to blend in with the existing remaining roof tiles. This is common practice, and one that may even be stipulated in the planning permission on properties in sensitive locations, such as areas of historic interest. For the remaining parts of the roof that require new tiles to be 'in keeping', you will need to source second-hand tiles from elsewhere.

If you are able, and have decided, to replace the roof tiles completely with a different type, it is possible that your local roof tile supplier will buy your salvaged tiles from you; more about this in *Roof tiles*, overleaf. Whatever the types of roof covering your project involves, make sure that if you are using a roofing contractor to carry out the stripping and renewing of tiles or slates, the possibility of salvaging material is discussed. If your project involves roof coverings such as Welsh slate (which is very expensive to buy), it is worth speaking to your local supplier who may be prepared to visit your property to advise you.

Building a new house

Even if you are building a new house on open ground there may be uses for the earth that needs to be removed for creating the foundations. For example, consider the work that you will be doing at the end of the project with regard to landscaping. You may want to build up some areas with general earth and then cover it in topsoil for planting, or you may be planning a raised garden wall, which will need filling with earth before laying the planting soil on the top. Obviously this will depend on whether you have enough room to keep it; if it means that by trying to save material you end up creating other problems, then removing the spoil/earth is a better option.

Refurbishment

When it comes to refurbishment, there may be a treasure trove of salvaged material that can be re-used or sold on to architectural salvaged material suppliers. Or you may need to find something that is no longer manufactured: you may be able to buy it at a specialist second-hand outlet.

You will find all types of specialist suppliers listed on the Internet who buy and sell salvaged material. Since the growth of the eBay phenomenon, you might also consider selling your salvaged items on-line, or use this website to find the item that you need.

As refurbishment covers such a wide range of properties and material, below are some of the most common elements, and also some that are worth looking out for on older properties.

Roof tiles

As stated under *Loft conversions*, you could look into whether your local roof tile supplier will buy your salvaged tiles from you. This is a common practice as many people will be looking for tiles that have an element of ageing or are weathered when trying to match in some new work. In this event, you would normally be expected to get the tiles down to ground level, and have them stacked on a pallet so that the buyer can easily load them onto his vehicle. However, it would be wise to make the necessary enquiries and weigh up all the costs involved before taking the trouble to remove and get them down to the ground unbroken.

Initially, though, you may not find a prospective buyer for your second-hand roof tiles; he may already be inundated. However, the buyer may give you an indication of how much he is prepared to pay for them when he has the space. If you are able to store the tiles yourself, you may make more money by selling them on eBay or by contacting your local roofing contractors or builders.

Guttering

If you live in a property that features traditional characteristics (such as cast-iron guttering and down pipes), planning conditions may require you to retain them. If this is not a requirement but you prefer to keep the appearance traditional, salvaging the original guttering could make the property more desirable to potential buyers at a later date. Cast-iron guttering can appear 'flaked' and beyond repair, but provided it is not cracked or broken it is easily cleaned and very adaptable to work with.

Windows

Replacing windows will require some investigation on your part as to whether planning permission is required. The original windows from an old property may not be salvageable, but this does not mean that you should just throw them away. There are certain parts of older windows that may be of use to others, such as patterned leaded light panels; any that are in good condition may be worth a considerable amount of money and so it is worth taking photographs of them and trying to locate a buyer. Alternatively, if you have permission to keep the traditional type of window and can re-use the old glass, this will retain the original characteristics, which is very desirable; in some cases, the glass can be irreplaceable due to the period and manner in which it was made.

Doors

In older properties (where you are having a complete change and removing the door frames that still have the original doors in place) it is worth considering salvaging them for re-use. Period doors are not cheap and you may not be able to buy the same type of door 'off the shelf' to match existing ones. If you are planning to carry out extensive refurbishment to a property where the doors have been changed to a type that do not match the period or style of the house, you may be able to find the correct type of door at an architectural reclamation store.

Door and window furniture

Apart from worn hinges and mortise locks that are clearly beyond economic repair, some door knobs, rim locks, window catches and window stays may cost a small fortune to replace. Where doors have been painted over many times, it is likely that the door furniture has been painted too. It may seem that they are totally worthless, and when you have been looking at them for years, you may not give them a second thought. However, under the paint you may find that the casement of the rim locks and door knobs is made of solid brass. Similarly with window furniture, particularly box sash windows, as most of the fittings were usually made of brass. There are chemical solutions that will remove layers of paint and will cost much less than replacing them.

Fireplaces

When you undertake a refurbishment project with a view to modernising the property with central heating and removing gas or open fires, it is worth considering retaining (or opening up) existing fireplaces. Traditional cast-iron fireplaces are now fashionable and aside from looking attractive, make the property more desirable to potential buyers. Alternatively, if you really do not want to keep the fireplaces (or due to the nature of work you have fireplaces that are surplus to requirements), they may be quite valuable and worth trying to find a buyer. If you have a fireplace that is decorated with ceramic tiles and is going to be redundant, you should make an effort to find out the true value. In some cases these tiles alone can be worth a small fortune.

Needless to say, the more ornate the fireplace, the more likely it is that it will be valuable. Fireplaces change hands for hundreds or even thousands of pounds on a regular basis, and latterly they are becoming harder to locate. You can buy new cast-iron fireplaces and surrounds, but the originals are always in great demand.

Radiators

Modern radiators are not really worth salvaging unless they have been specially made to fit a specific shape (for example a bay window); however, as in the case of the cast-iron fireplace, the cast-iron radiator is worth salvaging. As touched on earlier, many people are deciding to return traditional properties back to their original appearance where this does not cause practical problems, and where the building regulations are still being complied with.

Skirting and architrave

If you are carrying out extensive refurbishment requiring the removal of skirting and architrave, in this case you need to think carefully before salvaging it, as it can prove to be a false economy. Having been painted many times, they may need a great deal of work in order to improve the quality, particularly given the fact that their removal will have caused some damage. However, if, say, the skirting is extremely high and has a moulding that is not available 'off the shelf', and it forms part of the original feature that you want to keep, then it would be worth the time and effort. There is always a high level of satisfaction in re-using existing original material, particularly for the real traditionalist.

Sanitary ware

There is always a question mark over salvaging sanitary ware, due to the nature of its use. However, when it comes to items such as original cast-iron baths, there are specialists who can carry out work to bring them back to their original state. If you prefer not to keep an ornate cast-iron bath, it may be worth salvaging for selling. As far as re-using WCs and basins is concerned, unless they are in very good condition it is worth buying new ones that are in keeping with the style that you are aiming for. High-level cast-iron flushing units with a pull chain are popular in older properties, and it is worth salvaging these even if you are not going to re-use them.

External works

If you are having extensive external works carried out, there may be material that can be re-used that at first would not appear worth saving, for example:

- cast-iron manhole covers
- existing brickwork
- plants and bushes
- fencing
- topsoil.

There will always be some work involved in salvaging any material, and this needs to be balanced against the cost saving and how the finished work will look. Reproduction materials can be used if a 'period look' is required, but it is satisfying to use the original material where possible.

Remember that topsoil is not cheap, so if you are planning to re-design a large area (or if your site strip includes removing topsoil), you may want to put it to one side for re-use or sell it to a third party. This may mean that they won't expect to pay you for taking it off your hands, but as the cost of having material disposed of is not cheap, it is worth considering.

Demolition

The very nature of demolition work obviously requires that a specialist contractor must carry it out; and you should instruct the team to include the price of salvaging in their quote if this is financially viable for you. There will be a variety of materials that could be salvaged from a demolition project. Even if you do not intend to re-use any of the material from the demolition, there are salvage companies who will either buy some of the material or take it off your hands at no cost, which is worth considering. Some of the most common materials that are salvaged from projects involving the demolition of traditionally built houses include:

- roof tiles
- floorboards
- floor joists
- fireplaces
- leaded light window panes
- doors
- cast-iron baths
- bricks (subject to how they clean up).

Other elements such as floorboards and joists need to be removed safely as the building is demolished.

Material storage and protection

We have mentioned in various sections of this book the importance of material storage and protection. If this is not adequately controlled, replacement material will soon start to affect your budget. Even if you have plenty of room on your site for storing material until it is to be used, there are still some fundamental issues to consider. The following is a list of building material for which specific planning for protection and storage is needed:

- Bricks and blocks
- Sand
- Cement
- Timber – sawn
- Timber – prepared
- Roof tiles

- Plasterboard
- Plaster
- Fixings

Bricks and blocks

Unless a forklift truck is available, bricks and blocks are usually unloaded directly from the lorry by its own lifting equipment. If the bricks and blocks have to be moved by wheelbarrow, it is important that they are handled with care. If bricks are thrown carelessly into a wheelbarrow the 'face' side of the brick is likely to become chipped and the damage will be noticeable on the finished work. Unless blocks are going to be left 'fair faced' (natural block finish and not plastered) slight damage will not be an issue. However, lightweight blocks will break more easily, and heavy blocks will cause more damage to hands if they are not handled with care.

Bricks and blocks need to be covered at all times until they are laid. Most people assume that these are waterproof as they form the very fabric of a house; in fact, they are porous and when delivered to site in packs they are generally wrapped in plastic. As soon as they have been opened and placed around the site ready for laying, they are exposed to the elements. If they get wet it can cause problems for the bricklayers, particularly if they become waterlogged: the water will run out of the brick when it is laid, causing the wet cement to run down the face of the bricks or blocks, resulting in stains which are very hard to remove. Even after bricks are laid they may need to be covered with protective material, particularly if there is a chance that they will be affected by frost.

Sand

Loose sand also needs to be protected from elements such as frost and rain. Overnight frost can cause the sand to freeze which may delay the start of daily work. If the sand is not covered and it rains hard, it can wash away in a very short space of time. This can cause other problems, such as blocked drains, and it may also take time to clear sand from roads.

Cement

Cement is delivered in bags that have a thin layer of plastic to keep out the very light moisture content in the air. Bags of cement should be opened only when they are needed. Cement has a limited shelf life and should be issued with a date stamp on the bag giving the date of manufacture. Its shelf life is dependent on how it is stored; where possible, cement should be kept in well-ventilated and dry conditions. If you do need to keep bags outside, it is important to keep them protected with waterproof coverings, particularly at night.

Timber – sawn

Timber has particular characteristics. If they are not taken into consideration, the timber can warp or twist to such a degree that it may become unusable for the purpose for

Long floor joists, shown here, would be unusable if not stored appropriately prior to use, and should be temporarily braced during the construction phase

which it was intended. It may be delivered to site in different states depending on whether it has been treated with preservatives. If the timber has recently been dipped or impregnated with preservative, it may be saturated. This can make the timber very heavy, and it may be subject to movement as it dries out. Weather conditions can severely affect timber. If it is left in the sun, it can dry out too quickly. In extreme sunlight it can warp severely, causing large cracks, and this can also affect its structural integrity. Where possible, timber should be kept out of direct sunlight and covered. It is advisable to lay the timber on bearers on level ground, with bearers or strips of wood between each layer. This will allow it to dry out sufficiently before use.

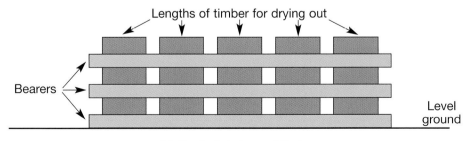

Method of drying out timber

Timber – prepared

As with sawn timber, prepared (planed) timber will twist and warp if it is not protected and properly stacked. Prepared timber (such as skirting and architrave) is usually thinner than sawn timber (for such as joists and studwork), and can lose its shape more quickly when exposed to even slight temperature changes. Prepared timber for internal use should be stored inside a well-ventilated area as close to room temperature as possible. In order to minimise warping or twisting, it should be fixed into position as soon as practical.

Roof tiles

As with bricks, roof tiles need to be treated carefully when being moved in order to avoid damage. Your supplier should advise you on any particular methods of stacking or storage.

Plasterboard

Plasterboard is very susceptible to moisture absorption and can become unusable if not stacked properly. Plasterboard needs to be stood on its edge against a solid wall at a safe angle that will not allow it to fall, but not so much that it will bow. Timber uprights (or a sheet of board material the same size as the plasterboard) can be used to stop the plasterboard from bowing. Unless plasterboard is kept inside at room temperature it will need to be covered to keep out moisture.

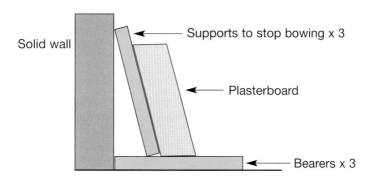

Method of stacking plasterboard

Plaster

As with cement, plaster has a manufacture date stamp on the bag and has a limited shelf life. Plaster is much more susceptible to moisture absorption than cement and the bags do not have a plastic layer to keep out low levels of moisture. The shelf life of the plaster will be dependent on the conditions in which it is kept. If you will be ordering large quantities, it could be worth checking with the supplier as to approximate shelf life of the particular product that you intend to use. It is advisable that, wherever possible, plaster is kept in very dry conditions and that not too much at a time is ordered if the weather is wet.

Fixings

Most fixings (such as nails and screws) are delivered in cardboard boxes, which, when wet, fall apart and spill their contents. It is advisable to buy or make (from plywood or similar material), a box with compartments to put the different-sized fixings in. Small boxes of screws and nails are easy to lose or can be easily stolen, or taken by sub-contractors by mistake. Door furniture and similar items should be boxed up and labelled for each door, then issued to the person who will be fixing them. If you allow tradesmen to rifle through boxes looking for locks or handles, parts and components will soon be damaged or lost.

Alternative material

If you place an order for material and find that for some reason it is not available when you require it, you may need to consider using an alternative to maintain continuity. In itself this is not a problem, provided that the alternative material meets the standards of the original that was proposed. Where the material has been specified by an engineer or architect, you may need to clarify with them that your alternative brand is acceptable. It may well be that the building regulations will not allow you to use a different type if it does not meet the right performance criteria. The Building Control Officer may be able to confirm whether the alternative material is acceptable.

If you do use alternative material and it does not meet the building regulations or engineer's specification, you may be required to remove it and replace it with the original specified material. This could obviously become an expensive operation in terms of unnecessary work and wasted material and time.

There are many types of material that are very similar so it would take a trained person to know the difference. An example of this is timber which has been graded for its strength in order to meet structural requirements. When timber is graded for specific purposes, it is stamped with the appropriate grading stamp to ensure that the carpenters know which material to use. Problems can arise here if the carpenters are not made aware of what they need to use and where to use it. If fixed in the wrong place, the material may have to be removed or strengthened with additional timbers.

The grading stamp, specified by the architect or engineer, which the Building Control Officer would expect to see on timbers

As there are many instances where alternative material is used, this is an area that needs close attention.

Safety, health and welfare

Your obligations as a domestic client

When employing the services of a builder or contractor, you are undertaking a health & safety obligation amounting to a moral rather than a legal obligation. This is, of course, if your project is one in which you intend living as your residential home. A domestic client is someone who lives (or will live) in the premises where the work is carried out. The premises must not relate to any trade, business or other undertaking.

Whether you are overseeing the project yourself or employing a contractor to carry out the work, understanding the importance of safety, health and welfare is very important. There are grey areas of health & safety regarding projects that are undertaken by homeowners themselves. Whereas the Health & Safety at Work Act 1974 (HASAWA) encompasses all work carried out by professionals during their daily activities, homeowners who are undertaking the work themselves are not covered by it. It must be stressed, however, that there is a moral duty of care by the homeowner to any person who has contact with the building project. Any person who has contact with the project has a right under health & safety law to be protected from danger.

If you are going to employ the services of individual contractors to carry out elements of the work (for example, bricklaying), they should have public liability insurance and work to good building practices. All appropriate measures should be taken to remove or reduce the risks of accident or incidents, by introducing methods of controlling the risk. When you consider employing the services of individual contractors or a main contractor, it is important to satisfy yourself (as far as is reasonable) that they are competent to carry out the work in a safe manner.

If you are responsible for the site and a person is injured due to negligence on your part, legal action could be taken against you. If you are in any doubt as to where you stand with regard to health & safety, or if you require any advice or information,

---

THE HEALTH & SAFETY AT WORK ACT 1974, OVERVIEW

Under the Health & Safety at Work Act 1974, employers have statutory obligations to adhere to, based on Common Law principles. The effect of the Act has been to bring ALL people at work (and others) under the protection of the law. The Act covers all employment activities and applies to employers, self-employed persons, subcontractors, visitors to places of employment, employees, directors and managers, members of the public, designers, suppliers, etc. It also provides the HSE with various enforcement powers.

contact the Health & Safety Executive helpline on **0845 3450055**, or visit the HSE website: **www.hse.gov.uk**

Safety

Safety is the most important issue to consider when undertaking any type of building work. You need first to consider what risks are involved in the processes of the work to be undertaken. Whether it is building a small wall, demolishing a building, or building a new house, safety has a major part to play in the planning, record keeping, and financial aspects of the job. Safety procedures and methods of carrying out the work safely can cost far more than the material and labour content of the work itself, so it is important to consider the safety implications before undertaking any task.

Put simply, safety (in the context of health & safety at work) is an area that requires us to consider the control measures that we should put in place before carrying out a specific task. As health & safety is an industry in itself that covers all work activities in all industries, you need to keep in perspective the actual amount of safety precautions and procedures required for your project.

Acrow supports for safe temporary propping

As you will see under the heading *Risk Assessments* (Chapter 7), there are ways in which you can identify the nature of risks associated with your project, and the potential for an accident to happen. Because of the seriousness of any accidents and the repercussions they can have, you need to make the issue a priority and ensure that not only do you implement safety precautions but that contractors do as well.

It can be tempting to cut corners for carrying out some tasks, for example where hiring a mobile tower would be a better option than balancing some boards on oil drums, or using 240 volt power tools instead of providing 110 volt power, or the correct tools for the job. Apart from the fact that introducing safety precautions and safe working practices may save lives, they will actually ensure that the work is carried out to a higher standard. Statistics collected by the HSE and other organisations prove that in the long term, higher health & safety standards contribute to higher efficiency levels, which in turn help to achieve set budgets.

It is worth investing in literature that details health & safety in construction; there are some useful guides to safety awareness that will help you to build up a better understanding of what to look for.

To give you an insight into the effects of accidents, the following information will highlight the seriousness of ignoring the issues.

Health & safety facts in Great Britain

- Falling from height or being struck by a vehicle cause the most fatalities.
- Slips, trips and falls on the same level cause the most non-fatal injuries to employees.
- Injuries sustained while handling, lifting or carrying are the most common kind of injury to employees resulting in three or more days' absence.
- Accidents involving electricity are very serious; about 1 in 30 of all electrical accidents are fatal compared with 1 in 600 of other types of accidents.
- Injury to workers in the first six months is over twice that in workers who have been with their employers for at least a year.
- Around 25 million working days are lost each year as a result of work-related accidents or ill health.
- Over 25,000 people are forced to give up work every year as a result of work-related accidents and ill health.
- Workers in small manufacturing firms are twice as likely to be killed at work as workers in larger firms in the same sector.

Accident-prevention awareness

When considering accident prevention, the following considerations should be taken into account:

- moral
- legal
- economic.

Moral considerations: Management has a moral duty for the safety and well-being of employees and non-employees, and to encourage safe working practices at all times.

Legal considerations: The Health & Safety At Work Act 1974 (HASAWA) places a general duty on employers for the health, safety and welfare of employees. Possible consequences of failing to comply with H & S legislation, approved codes of practice, guidance notes and accepted standards, will have a negative impact on organisations. Loss may result from the preventative (enforcement notices), punitive (criminal sanctions) and compensatory effects of law.

Economic considerations: Considerations should include the financial impact on the organisation of the cost of accidents, the effect on insurance premiums, and possible loss of service/production, corporate reputation and overall profitability. Costs may be direct or indirect.

The cost of accidents

This can be measured in terms of human, financial and economic costs to the nation. An accident which is apparently minor can have a major impact on both the company and the victim, when all related factors are considered.

Human aspects – cost to the victim

- Mental strain
- Suffering
- Loss of earnings
- Extra expenditure (medication, transport)
- Possibility of continuing disability
- Incapacity for some kinds of work
- Loss of leisure activities
- Possible loss of life
- Effect on family, friends and colleagues

Financial aspects – cost to the company

- Loss of skilled and experienced workers
- Loss of production
- Loss of profit from skilled workers
- Expense of re-training injured worker or training replacements
- Time lost by knock-on effect on other workers
- Loss of income from damage to corporate reputation
- Increased insurance premiums

An accident is commonly defined as:

- an unplanned event, which may or may not result in injury or damage. (A near-miss is by definition an accident, and should be regarded as a warning that a problem exists and that some positive action is required.)

Accidents are caused mainly by:

- human error – 88 per cent
- mechanical failure – 10 per cent

Those that are considered inexplicable account for 2 per cent.

Unsafe acts ...

- lack of understanding, training and supervision
- complacency and attitude
- not wearing PPE (personal protective equipment)
- procedures not known, not enforced, or ignored
- removal of guards and other safety devices
- poor communication

... cause unsafe conditions:

- trailing cables
- slip/trip hazards
- poor maintenance
- environmental: temperature, noise, dust, fumes
- workplace layout, falling objects.

Accident potential will be dramatically increased when unsafe acts and unsafe conditions occur at the same time.

Cost to the nation

- The expense/burden on medical or health services and facilities
- The burden on welfare benefits and social services provided by government

Known costs

If known costs equal, say, £1 (covering injury, ill health and damage), hidden costs may equal £8–£36 in terms of:

- damage to property, plant and equipment
- business/production interruption
- bad publicity
- sickness absence pay
- replacement staff pay

- loss of expertise/experience
- recruitment and training
- emergency supplies
- investigation costs
- clerical/administration costs
- legal costs and fines.

The true cost of injury and damage will only be assessed with any degree of accuracy as the company's information in this respect becomes more detailed and precise.

Health

Most people would associate health with being generally well. Health at work carries a different set of potential problems, particularly in the construction industry, although every industry has its own set of unique health issues in relation to the employees; we can regard being 'healthy' as being fit and able to perform a task. When risk assessments are carried out they highlight the areas related to maintaining the working environment to a safe standard. Anything that has the potential to cause a health problem should be assessed and the appropriate control measures implemented to avoid the risk becoming a reality.

In practice, we are all susceptible to taking risks. We are usually focused on a job, and when a situation arises that entails getting additional help, we may try just that little bit harder to do the job alone. For example, we may have to lift something that does not involve taking chances with safety but may have implications to our physical health. There is a fine line between 'health' and 'safety', illustrating why these words are usually referred to in conjunction.

As the manager of a project it is important that you not only consider your own actions, but that you are also aware of the actions of others. When it comes to 'manual handling', we may assume that a person who is of roughly the same build as ourselves is capable of lifting the same weight, but everybody has their own levels of strength. It is therefore very important to take note of the guidelines for manual handling and other health issues that affect groups of people.

When issuing any sort of instruction, you need to take various issues into account to avoid placing a person at risk, particularly if they are young or inexperienced.

- Is the person physically able to carry out the work?
- Is the person mentally able to carry out the work?
- Is the material hazardous?
- What PPE (personal protective equipment) is needed?
- What methods do we need to introduce to protect others?
- Do we need specialist contractors?
- Will other work affect the operation?

As you become familiar with individuals' strengths, weaknesses and levels of experience, you will consider the above as a matter of course. They are simple control

It is important that personal protection equipment (PPE) is used appropriately

measures that will help to maintain the health & safety aspect of your project, and when these questions are used in conjunction with risk assessments, your risk limitation will be noticeable.

One of the most important elements of health & safety to pay attention to is the use of personal protective equipment (PPE). It will enable those using it to work more efficiently and safely and you will find that in most cases PPE is inexpensive. The use of even the most basic equipment is all too often ignored in the domestic side of the construction industry.

> **The commercial side** of the industry is regulated by specific requirements before a project has even started, for example the health & safety policy of a building company, which spells out their arrangements for dealing with health & safety implementation and management.

It is important to use the right type of PPE. Your local plant hire shop will provide some of this equipment as part of their service. (However, if you are well organised you can make a saving by purchasing equipment elsewhere, as this may actually cost less than hiring.) Manufacturer's equipment information will usually detail the correct type of PPE to be used in conjunction with it.

PPE that you would expect to see on a construction site of any nature could include:

- hard hat
- goggles (specific for grinding or cutting with a disc cutter)
- ear defenders
- ear plugs
- masks (lightweight for general dusty work)
- masks (rubber with canister for gas emissions or paint fumes)
- gloves (heavy-duty cotton or rubber)
- gloves (latex)
- gauntlets (welding)
- knee protectors
- fire-resistant smock and trousers (welding)
- safety glasses (for general use with circular saws, etc.)
- steel toe-cap boots
- wellington boots (working with wet concrete)
- high-visibility jacket (for sites with vehicular or plant movement).

As you can see from this list, there are many items of PPE that you may require but it is unlikely that you will need them all.

Welfare

Workplace safety and welfare covers a wide range of issues. The following information will give an idea of the things that you can do and look out for when employing contractors. Contractors have a legal obligation to their workforce, and the following are just some of the issues that need serious consideration on all sites:

- Maintaining the internal working environment:
  - ventilation
  - heating
  - lighting
- Where their work involves machinery and motorised plant they have a duty to manage the movement of vehicles and pedestrians in the workplace.
- Preventing persons falling from heights.
- Preventing persons being struck by falling objects.
- Providing a safe workplace at all times.
- Maintaining the workplace and its equipment in a safe condition.
- Providing adequate welfare facilities.
- Ensuring safety when storing or stacking materials.

In general, the physical condition of a building site or construction work area is often a major contributing factor where accidents occur. The majority of accidents that occur in the workplace generally are due to slips, trips and falls. These accidents are easily prevented both in terms of time and expense. Working at heights needs to be very well controlled in order to avoid serious injuries or fatalities, and this also applies to areas where people work at or in the vicinity of vehicle movements. Employers, or individuals in control of building sites or construction work in general, should always carry out risk assessments on the work activities being undertaken to ensure that proper control measures are in place.

We can describe the provision of 'welfare facilities' as:

- adequate supply of wholesome drinking water
- adequate changing facilities
- adequate rest facilities
- adequate number of WCs and washing facilities.

Employers have an obligation to their workforce to provide specific welfare facilities during the course of their daily work activities. Due to the wide subject area and the numerous types of building work, it is only possible here to point out some of the important considerations.

If you are employing a builder to carry out the work, he should make arrangements to provide the necessary facilities for his workers. On a building site where there is sufficient space, the facilities should include as a minimum:

- provision of potable water
- site accommodation where the workers can:
  - change
  - dry out clothes
  - wash their hands
- facilities for eating food, drinking and resting such as:
  - table
  - chairs
- WC with:
  - hot and cold running water
  - soap
  - towels.

All facilities provided should be well lit and ventilated, with heating provided when appropriate. It is not always possible to provide these types of facilities on domestic projects due to constraints on space, or for any number of other reasons. However, every effort should be made to provide the appropriate facilities for workers to remain refreshed, clean and sanitised at all times.

General duties and obligations

All work activities that pose a high risk to employees (or other individuals) require a correspondingly high degree of effort to ensure that those risks are controlled. Similarly, those that pose a low risk should require a lesser degree of effort and time to control. Some duties are absolute, whereby an employer must comply. Others are qualified by terms such as 'so far as is reasonably practicable'; these standards will rely upon the courts for interpretation.

'So far as is reasonably practicable' means generally that the degree of risk must be balanced against the cost necessary to combat it. The employer (or his representative) must firstly identify and assess the risk, its severity, the frequency or duration of exposure to the risk and the number of people who could be affected. The cost of doing something about the risk then needs to be calculated. Cost is not only measured in monetary terms, but also in terms of time, effort and degree of difficulty. Most employers find that time represents the largest cost.

If it is not possible to remove the risk or hazard completely, it may still be reasonably practicable to control it. It is important to monitor and review the control measures regularly in order to ensure they remain reasonably practicable. The duty of care from employers to employees, employees to other employees and contractors to employers, etc., is spelt out in H & S and other legal documentation.

It is important to remember that it is your responsibility to ensure that whoever is working on your site is 'competent' to do the job for which they are being employed. The contractors themselves have the same duty of care.

The duty of care to visitors is covered under the Occupiers Liability Act 1957 and 1984; the level of protection necessary will be determined by the degree of risk. A visitor may become a trespasser should he enter any area where he is not permitted to go.

General duties of employers

There is a general duty under the 1974 Act which is placed on an employer to ensure 'so far as is reasonably practicable, the health, safety and welfare at work of all his employees'. The employer has to:

- provide and maintain plant and systems of work that are safe and without risks to health
- ensure that articles and substances are used, handled, stored and transported in a safe fashion
- ensure that such information, instruction, training and supervision is provided to ensure the health and safety of employees
- ensure that any place of work under the employer's control is maintained in a safe condition
- ensure that all means of access and egress are maintained and that they are safe and without risks to health
- ensure that a working environment is provided and maintained that is safe and without risks to health

- ensure that adequate facilities and welfare arrangements are provided for employees.

The Health & Safety (Information for Employees) Regulations 1989 require employers to either:

- display a poster containing basic information on health & safety to employees (entitled 'Health & Safety Law – What You Should Know')

or

- provide copies of the poster in leaflet form.

These general duties are wide ranging and ensure that the employer has a duty of care for employees whatever their work location. It is a legal requirement for a company employing five or more people to produce a health & safety policy statement. If you are employing a builder to carry out the work who is required to produce a health & safety policy statement, it would be worth requesting a copy. Seeing for yourself how the company proposes to implement their health & safety obligation may give you a clearer understanding of their professionalism. Employers with fewer than five employees do not have to produce a written safety policy statement, but that does not mean that they do not have to adhere to the HASAWA regulations.

Duty of employers to provide a written safety policy

The safety policy must contain:

- a statement of the employer's general policy with respect to health & safety
- the organisation and arrangements that exist for carrying out that policy.

The policy and any revision must be brought to the attention of all employees.

The health & safety policy statement consists of three parts:

- the general statement
- the arrangements that are in place to ensure that the policy is put into effect
- the organisation that exists to manage the health & safety.

The general statement should identify:

- the aims and objectives of the company.

The arrangements that are in place to ensure that the policy is put into effect should identify:

- the arrangements that exist for securing health & safety in the workplace
- details of what the employer is actually doing to secure health & safety standards in the workplace, such as providing:
  - appropriate welfare facilities

This is a sample of a general H & S statement, which includes some of the fundamental issues that you would expect to be addressed by a professional company:

The company recognises that health & safety has positive benefits to the client and organisation and is committed to achieve the highest possible standards. It also recognises that health & safety is continually changing and must implement good business practices to meet the changes. The approach will be to identify the risks and put in place the appropriate controls to ensure that our legal obligations are actively seen to be carried out.

The company will encourage a positive health & safety policy in its activities, which will be supported by senior management. The health & safety performance of both individuals and the company will be monitored to ensure that standards are maintained. Monitoring and reviewing of the health & safety policy will be carried out on a regular basis. In order to achieve and maintain our standards, the following will form the company aims and objectives:

1. The company will ensure the appropriate arrangements are made to plan, develop and review the policy statement.
2. Management will implement developed systems for the effective communication of health & safety matters throughout the company.
3. The company will provide the appropriate information, instruction and training to all employees, to ensure their competence with regard to health & safety.
4. The company will provide sufficient resources to ensure health & safety such as:
     i. Personnel
     ii. Time
     iii. Finance
     iv. Equipment.
5. The services of health & safety experts will be used appropriately, where the company does not have the skills or experience.
6. The company will ensure that adequate measures are taken for the health and safety of visitors.
7. The company will endeavour to ensure that all relevant statutes, regulations and codes of practice are complied with.
8. The minimum standards that the company will adopt are those required by law, although the company will always try to exceed these where possible.
9. The company recognises that all of its employees have a duty for the implementation of health & safety, and this is not just a function of management. Managers will have specific duties and responsibilities to ensure that the policy is implemented correctly. The company will ensure that manager's performance with regard to health & safety is monitored and recorded.
10. The company will ensure that accidents and 'near-misses' are fully investigated and appropriate action taken to reduce the likelihood of their occurrence.

> 11 The company will ensure that procedures and systems are implemented to ensure that safe equipment and plant are provided for employees and non-employees.
>
> The statement should be signed by the employer or managing director and dated. An organisation chart identifying the management structure of the company should accompany the safety policy statement, highlighting the key individuals who are responsible for safety, health and welfare.

- first aid provisions and trained personnel
- safety inspections and reports.

The organisation that exists to manage the health & safety should identify:

- all levels of management who have been assigned specific responsibilities which relate to their position and abilities.

Once the policy has been formally approved, the employer must bring it to the attention of employees. For the policy to be effective, the employer should take active steps to ensure that the employees implement what it states. This could include training sessions for employees. Anyone who has been identified in the policy as having key responsibilities (such as managers) should receive more in-depth training.

On a project of considerable size and duration, a comprehensive plan should be put together to deal with health & safety issues. Whether you are employing a builder, carrying out the work yourself or employing individual contractors and overseeing the work, it is important to provide a safe working environment so far as is reasonably practicable.

General obligations of employees while at work

The employee also has the following responsibilities in terms of health & safety.

- To take reasonable care for the health & safety of himself and of other persons who may be affected by his acts or omissions at work.
- To co-operate with the employer so far as is necessary to enable the duty or requirement to be performed or complied with.
- Not to interfere with or misuse anything provided in the interest of health, safety or welfare.

Employees have a duty under the Act and could, in theory, be prosecuted for their own injuries. The phrase *'acts or omissions'* means that the employee is responsible for the things he does (e.g. creating a hazard) and for the things he doesn't do, such as not removing or reporting a hazard he has found.

CHECKLIST OF H & S DOCUMENTATION AND CONSIDERATIONS

Policy

- Does the builder have a written H & S policy?
- Have you seen it?
- If the builder does not need one, what are his proposals for maintaining a safe site?

Signage

- Externally
  - Are pedestrians and neighbours adequately informed of potential dangers?
- On-site
  - Are sufficient signs placed in prominent positions for visitors and site personnel to comply with site rules and H & S regulations?
- Canteen:
  - Has a copy of the site rules been posted in the canteen?
  - Has good use been made of wall space in the canteen for signs or H & S information such as:
    - fire plan, emergency evacuation routes
    - arrangements for fire warning and procedures
    - equipment: types of extinguishers and positions
    - legally required poster containing basic information on health & safety to employees (entitled 'Health and Safety Law – What You Should Know')?
- Office
  - Has a poster been displayed containing basic information on health & safety to employees (entitled 'Health and Safety Law – What You Should Know')?

Personal Protective Equipment

- Provision
  - do you have the correct PPE?
- Use
  - Are you or are they using it?
- Storage
  - Are the right facilities available for its storage?

Site rules for:

- site personnel
- contractors
- visitors
- specific groups, e.g.:
  - delivery drivers
  - utility providers.

> **First aid**
> - Training
>   - Is there a trained first aider on-site?
> - Information
>   - Does everybody know where the first-aid equipment is and who the first aider is?
> - Equipment
>   - Is the appropriate equipment on-site?

Employees must therefore take positive steps to understand the hazards of the workplace and comply with the company's policy, rules and procedures relating to H & S.

Copies of certification of training

In order to satisfy yourself that a builder or particular individual has had the necessary training for the nature of work they are carrying out, you should request copies of certificates that they hold; this is because some activities, such as work that involves electricity and gas, are required to be carried out by contractors who have undergone specific training and who are qualified to certify their work formally.

If you are employing individual contractors who do not have any formal qualifications or proof of training, it does not mean that they are not capable in other fields. However, if you are in any doubt or if the individuals do not come well recommended, it is important to monitor their initial work in order to be satisfied that the standards you require are being met. In general, tradesmen who have undertaken specific training will have had elements of health & safety as part of their overall training. You may find that some individuals have undergone safety training on commercial building projects, and have a CSCS card (Construction Skills Certification Scheme).

The CSCS card is proof that the individual named on the card (similar to a credit card) is considered to be competent at his or her job, and also shows that the card holder has undergone health & safety awareness training or testing. The Government is committed to improving specific elements of the construction industry such as raising standards of both workmanship and health & safety. The aim is to have a fully qualified workforce by the year 2010 when it will be compulsory for workers to have proof of their skills and qualifications before they can work on-site. Many contractors are already implementing this and asking for all workers on their site to hold a CSCS card to prove their competence, and that they have health & safety awareness. Some clients are also specifying that builders and contractors will only be awarded contracts if they have a workforce with CSCS cards.

Induction and induction register

Induction procedures for all new employees or personnel working in an environment with which they are not familiar is common practice across all industries. As the construction industry has an environment with potentially more hazards, it is important that all site personnel are made aware of the site conditions and procedures before they start work. The bigger and more hazardous the site, the more comprehensive the induction needed. These are the main considerations to point out in the induction:

- Fire plan and procedures
- First-aid information and arrangements
- Security arrangements
- Specific hazards to be aware of
- Welfare facilities and arrangements
- Reporting procedures
- PPE requirements
- Site rules

It is obvious from the above that some of these points would require a written statement that the individual can take away and refer to as required. Once an individual has been inducted, their name should be added to a register (perhaps in the site diary) and their signature should be requested to confirm that they have understood what is required of them. This is an important consideration, so that you can be seen to have taken reasonably practicable steps to ensure that you are running a safe site and maintaining records.

General inspection

Builders and contractors who are required to maintain records and carry out inspections for specific elements at prescribed times should have the knowledge and training to fulfil that commitment. Elements that require specific safety inspections that need to be recorded include:

- working platforms
- excavations.

This does not mean that if you are running the project yourself you do not need to carry out inspections. You should, in fact, obtain the free leaflets and appropriate guidelines and information for inspection and reports from the Health & Safety Executive (HSE) website **www.hse.gov.uk/pubns/conindex.htm**, and make your contribution to ensuring that you are helping to provide a safe working environment.

If you do not have access to the Internet, you can order hard copies of the leaflets by calling the HSE on: **0845 3450055**. Many other free leaflets are available that will help you to run your project safely.

- General
  - Handling kerbs: reducing the risks of musculoskeletal disorders (MSDs)
  - Safe erection, use and dismantling of falsework (temporary support structure)
  - Cement
  - Noise in construction
  - Safety in excavations
  - Handling heavy building blocks
  - Inspections and reports
  - Personal protective equipment: safety helmets
  - Construction fire safety
  - Construction site transport safety: safe use of compact dumpers
  - Dust control on concrete cutting saws used in the construction industry
  - The Absolutely Essential Toolkit for the smaller construction contractor
- Chemicals
  - Solvents
  - Silica
- Working at Height
  - Inspecting fall arrest equipment made from webbing or rope
  - Working on roofs
  - Tower scaffolds
  - General access scaffolds and ladders
  - Preventing falls from boom-type mobile elevating work platforms
- Welfare
  - Construction (Health, Safety and Welfare) Regulations 1996
  - Welfare provision at transient construction sites
  - Provision of welfare facilities at fixed construction sites

Chapter 7

important things to know

Construction (Design and Management) Regulations 1994 (CDM)

If you are a client (other than a domestic client) for a construction project, the Construction Design and Management Regulations 1994 (CDM) will probably apply. The purpose of CDM is to ensure that health & safety is co-ordinated and managed throughout all stages of a construction project in order to help reduce accidents, ill-health and associated costs. Where CDM applies, you will have legal duties to discharge, which are enforceable in a court of law.

Your duties are the same whatever the size of project. However, the amount of information that you will need to provide will vary from project to project. Where any potential risks are low, little will be required of clients. Where they are higher, you will need to do more. It is important that what you do is proportionate to the risks and does not create any unnecessary paperwork.

When does CDM apply?

- CDM applies to all demolition and structural dismantling work, except where it is undertaken for a domestic client.
- It also applies to most construction projects.

There are a number of situations where CDM does not apply. These include:

- some small-scale projects which are exempt from some aspects of CDM
- construction work for domestic clients, although there are always duties on the designer, and the contractor should notify HSE where appropriate
- construction work carried out inside offices and shops, or similar premises, that does not interrupt the normal activities in the premises and is not separated from those activities
- the maintenance or removal of insulation on pipes, boilers or other parts of heating or water systems.

What are my duties as a client other than a domestic client?

As a client, you have to:

- Appoint a planning supervisor (either an individual or a company, e.g. a design team). The appointment should be made in sufficient time to allow the planning supervisor to develop a suitable pre-tender health & safety plan before arrangements are made for construction work.
- Ensure that the planning supervisor is provided with health & safety information about the premises or site where construction work is to be carried out. A planning supervisor has responsibility for co-ordinating the health & safety aspects of design and for ensuring that a pre-tender health & safety plan is prepared. Your designer may be able to advise you on this appointment.
- Appoint a principal contractor. Do this in sufficient time to allow the principal contractor to develop a suitable construction-phase health & safety plan before construction begins. A principal contractor has responsibility for co-ordinating health & safety aspects during the construction phase.
- Be reasonably satisfied that all those you appoint are competent and adequately resourced to carry out their health & safety responsibilities for the job in hand.
- Ensure, so far as is reasonably practicable, that a suitable construction-phase health & safety plan has been prepared by the principal contractor before construction begins.
- Take reasonable steps to ensure that the health & safety file you will be given at the end of the project is kept available for inspection by those considering future construction work. (The health & safety file is a record of information which tells you, and others, about the key health & safety risks that have to be managed during any future maintenance, repair, construction work or demolition on the building.)

Can I appoint someone else to carry out my duties?

If you wish you can appoint an agent to act on your behalf as client. If you do, you should ensure that they are competent to carry out your duties. If you appoint an agent, they should send a written declaration to the Health & Safety Executive (HSE).

It should:

- state that the agent is acting on your behalf
- give the name and address of the agent
- give the exact address of the construction site
- be signed by or on behalf of your agent.

Can I appoint myself to carry out other duties?

You can appoint yourself as planning supervisor and/or principal contractor provided you are competent and adequately resourced to comply with your health & safety responsibilities.

Where can I obtain further information?

If you have doubts about whether CDM applies, contact HSE's information line, tel: 0845 3450055, or write to HSE's Information Centre, Caerphilly Business Park, Caerphilly CF83 3GG.

Managing health & safety in construction

Construction (Design and Management) Regulations 1994. Approved Code of Practice and guidance. HSG224 HSE Books 2001 ISBN 0717621391, also provides further advice, as does the website: www.hsebooks.com

Risk assessments

Builders and contractors have a duty to comply with health & safety regulations as specified under the Health & Safety at Work Act 1974. As part of these regulations, risk assessments should be conducted to identify specific measures that need to be taken in order to protect their employees and all others having contact with the work.

Construction by its very nature involves activities that are dangerous if not approached in the right manner. For example, when dealing with excavations, working at height and during manual handling, we need to be sure that the risks are minimised. If you are planning on carrying out the work yourself, it is worth taking time to analyse some of the risks that are involved in your project. Builders and contractors should provide back-up paperwork and supportive material for the work that they perform. Unfortunately, in the domestic market this is not always carried out to the degree that it should be.

In order for you to understand some of the issues involved, here is a brief understanding of risk assessment and associated hazards. Rather than try to break this down to cover a domestic project where the regulations do not necessarily apply, it is considered that risk is an issue that should be taken seriously and as such should be viewed as it would normally be in a working environment.

Application

The MHSWR (Management of Health & Safety at Work Regulations 1999) apply to all industries and work places. They are based on the principle that the work to be undertaken is planned well in advance in order to identify and, where possible, remove hazards.

Where a hazard cannot be removed it is important to know the control methods that will be introduced to minimise the risks associated with it. Employees have a statutory duty to co-operate with the employer as far as possible in order to implement the specific actions.

Scope

There is no prescribed format against which a risk assessment is to be undertaken. Regulation 3 of the MHSWR states: '...a risk assessment must be carried out by a competent person for the purposes of identifying the measures that employees or self-employed must take to comply with their duties.'

> **If you are overseeing** the project yourself and employing labour-only contractors, it would be advisable to contact a health & safety consultant to ensure that the risks on your project are appropriately controlled.

To be suitable and sufficient an assessment must:

- correctly and accurately identify the risks
- quantify the risk (measure)
- prioritise the actions required
- take into account any existing control measures
- identify any legal requirements
- provide sufficient information on which to base any control measures necessary.

Interpretation

Hazard: the potential for something to cause harm, such as working on a ladder, or with machinery.

Risk: the likelihood that harm will actually occur from exposure to the hazard together with the likely injuries that will occur as a result and the likely numbers of people that this will affect.

Examples of a hazard:

1. Climbing up or down a ladder
2. Operating abrasive wheels
3. Handling chemical substances
4. Walking on uneven or slippery floors
5. Use of electrical equipment

In the above examples the risks are:

- falling
- contact with or explosion of the abrasive wheel
- exposure to chemicals
- slips, trips or falls
- electric shock or burns.

The extent of risk can only be quantified by looking at any existing control measures.

In point 1 under *Examples of a hazard* (above, regarding use of a ladder), the factors that greatly increase the risks are:

- ladders not secured properly
- damaged rungs
- incorrect length
- poor access to ladder
- wrong angle (it should be placed at a 75° angle, i.e. 1:4).

These risks can be effectively reduced by simple control methods, for example:

- by tying the ladders
- making regular checks for damage
- ensuring ladder is correct length for purpose of use
- ensuring that there is clear access
- ensuring that the correct angle is determined.

Risk Rating

The Risk Assessment should enable the employer to establish the degree and priority of risk. To enable this to be carried out a 'Risk Rating' can be used for each hazard, identified using the format below:

Hazard		Risk	
Major	3	High	3
Serious	2	Medium	2
Slight	1	Low	1

The Risk Rating is the hazard multiplied by the risk

Hazard	Risk	High 3	Medium 2	Low 1
Major 3		9	6	3
Serious 2		6	4	2
Slight 1		3	2	1

> **In order to** apply a rating properly, the appropriate amount of experience would be required to know the values to use. If you do not have sufficient experience, you would need to consult an expert.

Risk control

It is the duty of the employer to reduce the risks (as far as is reasonably practicable) by the use of control measures. These control measures need to be demonstrated in theory by the use of Risk Assessments and Method Statements (where required) prior to the work being carried out. Method Statements do not necessarily need to be produced for each action, provided that existing control measures are adequate.

Records

Risk Assessments and Method Statements must be:

- written in comprehensible form
- signed and dated by the person making the assessment.

Employees must be:

- informed
- instructed
- trained (where necessary) as to the risks involved, and the control measures needed to carry out the task.

> **In the event** that an accident or near miss occurs, resulting in an investigation, the first line of enquiry will require the employer to produce Risk Assessments and/or Method Statements previously carried out for that operation.

Risk Assessments should be carried out by qualified personnel, who understand the control measures that need to be implemented. It must be stressed that every project is unique and should be treated individually.

Method Statements

What is a Method Statement and what information does it contain?

- It is a document detailing high-risk operations.
- It is a descriptive account of the sequence in which an operation will be carried out.
- It lists the materials that will be used.
- It describes the location of the work.
- It details the tools or plant to be used.

- It details the safety precautions and personal protective equipment (PPE) that will be put in place to protect the worker and others.
- It is issued to competent persons who will carry out the work and is kept on file.

Method Statements are documents that are used extensively in major building projects and may be required under the CDM (Construction Design Management) regulations. If your project does not come under CDM, it does not necessarily mean that you do not need them. Method Statements are mainly used in the commercial sector and are supposed to be issued for all works that carry any element of risk. There is no reason why they should not be used within the domestic sector; individuals who use them will undoubtedly minimise the risk of an accident, and they will also help to identify costs and materials.

The following is an example of a Method Statement for working at height.

SAMPLE METHOD STATEMENT	
Circulation: A N Other	**Date:**
Site address:	
Details of work: Replacement of gutter and new timber to high level gable	
Location of work: Front and rear of property	
Plant to be used: 2 Lightweight towers, 110V drill, 110V saw	
Method to be adopted: Physical barriers will be placed in positions to stop anyone from wandering into danger areas. Warning signs will be placed around site warning of men working overhead. Towers to be erected by trained and experienced personnel on stable ground, and legs adjusted to suit any variations in ground level. Outriggers will be fixed to tower for additional support and stability. Existing cast-iron gutter to be removed and lowered to the ground by use of rope. Existing timber to gable to be removed and lowered to ground in the same manner as gutter. New guttering and timber to be passed up safely within the tower sections for fixing.	
Material to be used: 16m of 200mm x 25mm prepared timber, 30m of 150mm black plastic guttering, corner sections, connectors, brackets, 75mm screws, black japan screws, wood stain	
PPE to be used: Goggles, gloves, hard hats, boots	
Safety equipment to be used: Physical barriers and warning signs	
Prepared by:	

Contractor liaison and communication skills

We have looked at the formal procedures for communicating with the contractor and professional teams. It is important to understand that the construction industry attracts a wide variety of people who may not have the necessary skills to communicate effectively. The UK construction industry is made up of tradesmen of all nations, and it is not unusual to encounter communication problems as a direct result of language barriers. It is important that instructions are clearly understood, and it may be that additional supervision is required to ensure that instructions are carried out correctly.

If you are employing 'labour-only' contractors, it is important that you set in place a communication structure in which, for example, you can liaise with foremen of different trades. If you try to maintain control over each person on the site, you will be continuously issuing instructions. The number of foremen that you employ depends on the size of the project; however, having a person who specialises in particular trades would allow you to delegate. In general, the ratio for effective supervision of work on construction sites can vary according to the type of work being carried out. However, it would not be unusual for a foreman to supervise anything from four to ten site personnel. Once the figure rises to more than ten, things can become unmanageable. Where the work is very technical and requires the foreman to set out and check ongoing work, four to six subordinates would be enough to deal with.

As previously mentioned, it is very important to endorse any verbal instructions with written ones, and sketches where appropriate. It would be beneficial to have a duplicate pad at hand at all times, as it may not be necessary to issue formal instructions for things that require less significant detail. A duplicate pad is very useful in these situations, as copies of written information and sketches can be immediately issued to all concerned to clarify specific points or details; it is the small detail that can cause clients to become dissatisfied when things are not as they expected.

If you have any concerns whatsoever about the standard of work, health & safety, or any other matters, it is important to write to the contractor in the first instance. If there are concerns that need addressing immediately, stop the work completely and contact the contractor: he will soon make time for you if the workers are not productive.

Discipline on-site

If a situation arises where a worker is ignoring the site rules or acting in a dangerous manner, you should warn the person formally that you will be contacting their superior and that you wish them to stop whatever it is they are doing. It is important not to get into a confrontation with any site personnel if it can be avoided. Always keep disagreements or disputes low key, and where possible try to resolve them without the need for legal action. Where a contractor is obviously not capable of carrying out the work adequately, or does not have the personnel who can turn out the work to the standards required, you should immediately request a meeting. Before any more work is conducted, request a written proposal of how he intends to address the matter. If you are not fully satisfied with the proposals, you should terminate any agreement or contract that you have, and re-plan the work with an alternative contractor. These issues should be detailed in any contract or agreement that you have with the contractor.

Where you are employing 'labour-only' contractors, it is important that you have a pre-planned system of dealing with individuals who are not meeting their obligations. The type of problems that require you to have a disciplinary procedure in place include:

- ignoring health & safety issues
- not adhering to site rules
- poor time-keeping
- lack of output generally

- poor standards of work
- misuse of tools or machinery
- theft
- general lack of co-operation.

There are a number of methods for dealing with disciplinary procedures; however, it is important that site personnel know what these procedures are. If site personnel know they will be disciplined for not following set guidelines, they will be less likely to break them. An example of implementing disciplinary procedures would be as follows:

1. Give a verbal warning.
2. Give a written warning.
3. Dismissal.

Discretion is always the best way forward and formal warnings may seem unnecessary; however, if you do not have some guidelines and standards to work to it can prove difficult to maintain control.

There are situations where instant dismissal is necessary and, if the reasons are serious enough, appropriate action should be taken.

SELF CONTROL

The construction industry attracts a variety of individuals, and site managers can become exasperated by the actions and behaviour of some site personnel. In the event that you have to reprimand someone, ask them to leave the site for behaving in a dangerous manner, or not producing the standard of work that you expected, you may well meet with some resistance. Unfortunately, this is not uncommon, as many workers in the construction industry can be undisciplined.

The best way to deal with this type of situation is to try to remain calm and act in a professional manner. Acting unprofessionally could make matters worse. If you find that an individual is intimidating, it might be appropriate to liaise with a responsible and trustworthy member of the site team, who may be able to act on your behalf or support you in carrying out your actions.

Security measures

From a security and safety angle, there are many measures that need to be taken to protect the property from being entered by unauthorised persons. The area and type of work will dictate the level of security that is required, and below are some of the elements that you may need to consider.

Fencing/hoarding

There is a variety of fencing that can be hired which includes 2-metre-high panels made from wire mesh or mild steel corrugated sheets. These are supported by slotting into concrete or heavy-duty rubber block feet. The panels are fixed together with a clip that is bolted into place to give the fence rigidity. This type of fencing is very adaptable and easy to move when required; however, as it is lightweight and not fixed to the ground, the security aspect is minimal. This type of fencing is a more physical barrier that would take a little effort to move, but in general would be classed as a semi-secure fence.

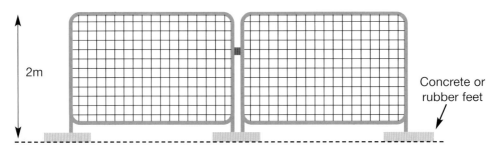

Typical mesh fencing

For a site that requires more security it would be advisable to build a more substantial fence or hoarding. On long-term projects (and where the fencing could remain in place without causing inconvenience), it would be more secure to place timber posts into the ground and fix plywood panels to them. Lockable gates would have to be constructed at the site entrance to allow easy access for deliveries, etc.

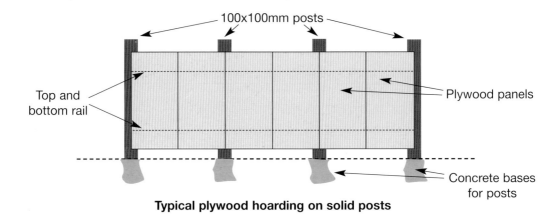

Typical plywood hoarding on solid posts

Warning signs

If your site is unoccupied and in an isolated area, it may be worth employing the services of a company who specialise in making security visits to sites on a regular basis. If you do this, you can affix signs to the fencing or building warning any potential trespassers that the site is monitored and that action will be taken against offenders.

Scaffolding

Scaffolding can be alarmed so that when not in use, any unauthorised access will set off an alarm and (if necessary) trigger lighting. These systems are very good for bringing peace of mind, but they can often be set off by cats or other animals or the movement of tree branches, which can be annoying. As a minimum security measure, all access ladders to the first level of the scaffold should either be removed at the end of each day, or adapted so that they cannot be climbed. Whatever project you are undertaking, it is always advisable (where appropriate) to keep windows closed where scaffolding has been erected.

Plant and equipment

Where there is expensive plant and equipment, there will often be the possibility that thieves may well operate. It is not always practical to remove equipment from site on a daily basis, and the only way to safeguard it from being stolen is the provision of security measures. A solid fence or hoarding with securely locked gates will make it difficult for opportunistic thieves. Most modern motorised plant incorporates good immobiliser systems; however, this would not deter a thief who has the equipment to remove plant by other means. There are other steps that would make it more difficult for determined thieves, including:

- heavy-duty chains and padlocks
- blocking/trapping smaller plant with heavy plant
- locking smaller items of plant in the site container or office.

You may consider that providing a site container is too expensive; however, you may be surprised at how much it can save by keeping materials and plant locked away.

Access to neighbouring properties

An area that needs to be given a lot of consideration is neighbouring properties. Because construction work involves many different aspects that could affect these, a risk assessment should be carried out to judge the effect that the work will have on the neighbours. Where scaffold has been erected between two houses or close to adjoining properties, warnings should be given to neighbours that while you will make every effort to keep the scaffold secure, they should keep their windows closed. Remember that while a scaffold is a means to carry out work, it is also a means of access to an opportunistic thief.

Site rules

You will need to consider very carefully what your site rules will be. If you are employing a builder, they should have site rules that the workforce are expected to adhere to, and you will need to know what they are as you may wish to add to them.

During your project you may need to alter the scaffold two or three times for the different trades to carry out their work, as shown here where it is being adapted for the roofers

If you are setting your own rules, there are some fundamental issues that you should consider, for example:

- Smoking
  - A no-smoking policy should be introduced on-site.
  - A smoking area could be made available.
- Use of WC
  - Which WC is to be used and who is responsible for maintaining it?
- Conduct of workers
  - Conduct of workers to be good at all times.
- Signing in and out of site each day
  - All site operatives and sub-contractors should sign in and out each day, so that in an emergency a head count can be carried out.
- Use of private equipment
  - If you have equipment that is not for general use, this should be clarified.
- Voltage of tools
  - For safety reasons only 110v tools should be used.
  - 240v tools should be excluded from site.
- Radios on-site
  - If radios are allowed, volume levels should be appropriately set.

- Parking on-site
  - If parking space is limited, indicate who can and cannot park on-site.
- Clearing of waste and debris as it is produced
  - As the project progresses, waste and debris will accumulate. It is important to point out who is responsible for removing it and where it is to be removed to.
- Dress code
  - It may be a requirement that site personnel wear shirts or high visibility vests at all times, particularly where there are motorised machines on-site.
- Alcohol
  - It is advisable to exclude the taking of alcohol or drugs during working hours.
- Access
  - It should be stressed that access and egress to all areas should be maintained at all times, including for emergency vehicles.

Accident prevention

In order to operate a safe site, it is important that all personnel understand that they each have a part to play. By encouraging the site team to adopt a safe approach to their work, they will be contributing to overall safety on-site, and the following are some of the fundamental elements:

- Understanding what a safe system of work is
- Following a safe system of work
- Appropriate training or self education for understanding how to recognise safe systems of work
- Correct and appropriate use of tools
- Appropriate training where required
- Providing appropriate supervision
- Providing and using the correct PPE (Personal Protective Equipment)
- Providing appropriate safety signs
- Adhering to site rules
- Recognising weaknesses and taking action

Staying focused on how the site operates, and understanding that a small sequence of events can combine to cause serious problems or accidents, will help to maintain a safe working environment. Taking an occasional step back and concentrating, for example, on health & safety issues may prevent poor attitudes developing on-site, which may result in dangerous working environments.

Safety awareness

There are many inherent health & safety issues with which the construction industry has specific problems. For example:

- Noisy work
- Dusty work
- Work involving naked flames
- Excavations
- Lifting and handling
  - Carrying material
  - Lifting awkward loads
  - Setting material down
- Working in confined spaces
- Working at height
  - Stepladders
  - Roof ladders
  - Scaffolds
  - Mobile towers
  - Mobile elevating working platforms
- Working with or near electricity
  - Electricity below ground (buried cables)
  - On-site cables
  - Overhead cables
- Working with hand-held power tools
  - Drills
  - Petrol-driven tools
  - Circular saws
  - Chain saws
  - Angle grinders
- Working with or near mobile and static plant
  - Forklift truck
  - Dumper
  - Excavator
  - Mixer
  - Compressor
- Working with hazardous substances, which could include:
  - Toxic or very toxic
  - Harmful
  - Irritant
  - Corrosive

There is an abundance of safety-awareness information, and it is the site manager's duty to ensure that all site personnel know their own responsibilities. If you are in any doubt about the ability of yourself or any person on-site regarding safety awareness, you should make arrangements to obtain the appropriate information or training.

Personal Protective Equipment should always be available and used where appropriate, though initially every effort should be made to remove any risks. If risks cannot be removed, they need to be controlled.

Prioritising

It is not always easy to deal with the day-to-day activities on-site and occasionally you may be forced to prioritise. If you do not prioritise when demanding situations arise, you could be in danger of losing control of the situation. This is also worth taking into account if you are employing a builder to carry out the work. If some work appears to be at a standstill or falling behind, do not be tempted to allocate personnel from another part of the project – this will just slow *everything* down. Sometimes it is better to get at least some areas of the project completed so that other trades can continue in the sequence, rather than having a great many areas waiting to be finished.

Delegating responsibility

If you are employing individual contractors to undertake specific elements of the work, it may be useful to delegate some of the supervising responsibility to those who have the maturity, experience of work, and knowledge of managing others. You may need to ask for proof that they have supervisory skills and then offer an incentive or bonus in order to delegate responsibility. If you are satisfied that the person is competent, and you have enough work to do yourself, you may find that it is worth the investment. Do not delegate responsibility to someone who is not able to prove that they can take the responsibility seriously.

Maintaining accountability

Although you may be able to delegate responsibility, it is not possible to delegate accountability if you are the person in overall charge. In order to manage a site properly and professionally, you will need to be personally accountable; also for the legalities of your position. Individual tradesmen are required to work to the HASAWA 1974 and they have a duty of care to themselves and others on the site. If you are the person in charge you have an overall duty of care to ensure their safety as well as your own.

Chapter 8

coming to an end

Good working practice and quality control

From the client's point of view, it is important to see evidence that the contractor has a set of core values by which they set their standards. Many companies pride themselves on the fact that, for example, they provide value for money, or have a good health & safety record. In order to satisfy yourself that you are employing a quality company, here are examples of some of the evidence that you should look for when carrying out reference checks or general observations:

- Does the company have a good reputation?
- Do they have a good health & safety record?
- Do they provide value for money?
- Do they complete projects on time?
- Do they have good management and reporting systems?
- Have they got sufficient labour resources to complete your project?
- Have they got the financial resources to finance your project up to valuation and payment stages?
- Have they completed work of a similar nature to the standards that the client specified?
- Do they have a professional attitude?
- Is their equipment and transport in keeping with the standards that they verbally express?
- Looking at their other sites, do they give you confidence?

In the first instance, any good builder will be seen to employ certain basic working practices that indicate the level of consideration shown towards the client, and the neighbours and residents. They will have systems in place to protect visitors to the site and people passing by, such as the placing of notices warning people to be cautious when approaching the site entrance, and signs showing visitors where to report before entering the working area. Of particular concern would be that they have given thought to people with special needs, such as those who are visually impaired, and this may include illuminating scaffolding or hoardings erected on the pavement. People who are hard of hearing or have difficulty with mobility may bene-fit from being directed to cross the road before passing the site, particularly where vehicles are entering and leaving on a regular basis. People in wheelchairs, as well as people pushing prams and pushchairs, would certainly appreciate being alerted to slippery or uneven ground, which can be caused by lorries and site vehicles frequently coming and going.

By its very nature, construction is usually noisy, and if a builder has put rules and systems in place to minimise noise levels, this is usually a good sign of professionalism. Where a contractor can demonstrate on paper the systems for dealing with site manage-ment, it is important that these systems are seen to be acted upon.

Builders should be able to demonstrate (on paper) how, for example, they would deal with:

- keeping footpaths and roads clean
- keeping the site clean
- disposing of waste and debris
- keeping dust and noise levels down
- taking delivery of materials
- storing material
- maintaining a safe site
- ensuring the safety of pedestrians
- maintaining site plant and equipment
- avoiding accidental pollution
- managing the project, including health & safety
- emergency procedures.

If you are in any doubt about how a builder or contractor is going to deal with a situa-tion, ask for proposals in writing. If you are not satisfied with the response, you then have the option to discuss how you would like to have specific issues dealt with, or take your business elsewhere.

Material storage and complaints procedure

In order to minimise the risks exceeding your costs, it is important to have a procedure to deal with material that is delivered damaged, of poor quality or is completely differ-ent from the order that was placed. Delivery drivers are not usually responsible for sorting or loading the material on their vehicles, and as such they do not check it for quality. It is the responsibility of the yard manager to ensure that the material they provide is sent out in good condition. The way in which material is stored will determine the condition in which it is sent out to the customer: for example, timber needs to be stored in well-ventilated areas.

All timber needs to be stacked carefully while at the suppliers. Timber will normally need to be laid flat on a level surface in order for it not to warp or twist. If left in the sun it will have a tendency to dry out very quickly and can twist or bend to such a degree that it becomes totally unusable. For this reason your own storage facilities need to be properly thought out. Your material may only be on-site for a couple of days before it is fixed in place; however, in extreme conditions such as frost, rain, or heat, the material could become unusable.

Dos and don'ts

Since this book is aimed at those who will be undertaking the work themselves, managing individual contractors or employing a builder to carry out the work, it is important to interpret the information correctly. This means that specific information may need to be applied in a way that suits both the project and the personnel involved. As a site manager (or client who is overseeing your own project), you will come into contact with a variety of people who have very different attitudes to their work environment, in particular health & safety. You will also meet people who are very articulate and some that cannot communicate effectively. Whatever your role, it is your money being spent and you need to maintain control at all times. In order to run and successfully complete a project from start to finish, it is important to remember some of the fundamentals outlined in this book, covered simply by the dos and don'ts.

Do:

- think carefully about what you want to build and whether it suits your needs
- plan your project with realistic timescales
- take time to research elements that you don't understand
- implement control procedures for the various elements of work
- consult experts where necessary
- check to see if you need planning permission
- obtain a sufficient amount of quotes
- compare quotes and be prepared to negotiate
- have all agreements in writing
- think about the terms of any contract in detail and confirm them in writing before signing a contract
- carry out appropriate reference checks for builders, contractors or individuals where necessary

- consider changing builders, contractors or individuals if they do not meet your standards
- set up your site management system for making and keeping records
- study and implement health & safety procedures
- obtain appropriate certificates, guarantees and warranties
- send off guarantees, etc., where required in order to validate them
- act professionally at all times
- remember that last-minute changes cost money
- consider contingency plans in general
- hold regular meetings, formal or informal
- expect to receive written reports of progress
- remember that if you fail to plan you are planning to fail.

Don't:
- be tempted to make quick decisions if you don't need to
- necessarily accept the cheapest quote
- accept poor standards of work or material
- take chances
- take risks with health & safety
- get into confrontations – remain calm and in control
- pay builders or contractors up front unless it is for material or equipment that needs to be manufactured off-site, for example:
  - windows and doors
  - kitchen units
  - wardrobes
  - specialist joinery work
- ignore environmental and local issues
- leave too much to memory (write it down)
- give instructions without knowing the financial and timescale implications
- forget that everything needs to be planned.

Tips and hints

Unfortunately, it is a fact that some builders may show genuine interest in quoting for a project, but then for a variety of reasons decide that they are not going to. Builders may not be prepared to divulge why they are not going to quote but, for example, it could be that:

- the job is too big
- the job is too small
- the job is too technical

- there are logistical problems that they are not happy with
- the job will be too difficult
- they may not want to enter into a contract
- they can't afford to finance the initial stages without upfront payments
- the job is a considerable distance from their usual geographical location
- they lack the necessary resources.

Whatever the situation, these are common occurrences that can leave the client very frustrated. One way of reducing the risk of wasted time, effort and cost is to ask any potential builder if there are likely to be any problems in providing a quote. You could produce a list of concerns that need to be addressed (as above) and, of course, there may be other issues that are of concern to you. When you are satisfied that the builders are genuinely interested and have given them all the appropriate information for pricing the job, it is important to set a date for receiving the quotes. You could also set a date for meeting each of the builders to discuss the quote in detail. Once the builder knows that you will expect to meet for a discussion, it is likely that he will put the quote together in more detail than he would for someone who just provides drawings and basic details without suggesting a meeting.

It can pay to advertise – displaying your advertising board could result in contacts from potential clients

Whatever size project you are undertaking (and however you are undertaking it), remember to take plenty of photographs. Every time you visit the property (or if you are there every day), get in the habit of taking photographs. If you do not have a digital camera, it is worth investing in one for the project. Photographs can be downloaded onto a disc at a very reasonable cost.

Look for the signs

If you are planning to employ the services of a builder, take time to look at a project that he is currently working on and how he prepares and presents the quote. Specific signs to look for include:

Site

- Is consideration given to neighbours, pedestrians and road users?
- Is the site tidy and well organised?
- Do the site personnel look presentable?
- Are the builders prepared to advertise themselves, for example with signs or boards placed on the scaffold or at the front of the address being worked on, or perhaps with advertising on site workers' clothing and on vans?

A well-organised and tidy site is an indication of good building practice

- Is their equipment clean and appropriate?
- Are there sufficient warning signs?
- Is the material stacked well and protected (where appropriate)?

Preparation of quote

- Did the builder take sufficient time to inspect specific parts of the property prior to quoting, and were appropriate questions asked to clarify any points?
- Did he point out that you could perhaps make savings by doing something differently, or using different materials?
- Did he present the quote professionally and with a covering letter?
- Did he offer you the opportunity to discuss the quote in detail?
- Did you feel confident that the builder wanted to do the job?

When the work commences, always be aware of the way in which site personnel work. As the job progresses, debris and waste material will accumulate, as is the nature of building work. It is not difficult to have any waste material or debris removed from the work area to a collection point. If construction is allowed to continue without areas being cleared, not only will it slow the work down but it will also cause health & safety risks. Ask yourself the following:

- Has someone been tasked with clearing waste and debris?
- Have appropriate methods been introduced to remove waste?
- Is the waste being bagged up?
- Have efforts been made to reduce dust?
- Is waste or debris being allowed to fall between floorboards?
- Is the scaffold being kept clear?
- If a skip cannot be used, is the waste/debris being removed on a daily basis?
- Are materials being kept where they are not causing a danger or getting damaged?
- Are the workers using appropriate PPE?

Share information

In order to maintain a good working environment, it is important that everybody works as a team. It does not matter whether you are employing a builder or individual contractors, everybody involved in the project should be informed of the specifics. For example:

- Site induction issues
  - Fire precautions and procedures
  - Welfare facilities
  - First-aid facilities and personnel
  - Emergency procedures

- Health & safety matters
- Site rules
- Procedures that need to be followed for carrying out additional work
- External considerations:
  - Neighbours
  - Pedestrians
  - Local businesses
  - Parking
  - Noise
  - Deliveries
  - Traffic

Apart from being prepared to share your information and having a method of conveying it effectively, it is also important to let all site personnel know that they have a duty to report specific information. For example:

- Any hazards they may find
- Unauthorised persons on-site
- Any security risks
- Dangerous work activities (by others)
- Risks to pedestrians
- General health & safety problems

Promoting a team approach will help to ensure that communication of appropriate information is delivered to the right people at the right time.

Looking forward!

Although you may be building for the present, it is worth considering what you might need in the future. By thinking ahead, you could put in place (for an additional but relatively small cost), some elements of building work that would save a considerable amount of disturbance to the building at a later date. For example, if you are currently building a single-storey extension, putting in deeper footings may allow you to build another storey at some point.

If you are planning a small extension (but have room for something larger and it is within your budget), it might be sensible to build bigger now as a greater incentive for potential buyers.

Do not leave it to the last minute to approach builders; once you have decided to build or extend, start putting your plan together and do some fact finding. The more information that you can gather early on about all of the services you will require and people you will need to contact, the less likely it will be that you make ill-informed decisions.

Keeping everything in an orderly fashion is of benefit both to the builders and the client

Learn from the past

If you have had good (or bad) experiences in the past when dealing with builders, contractors or individual tradesmen, remember what worked well and where things could have gone better. It may be useful to speak to others who have had work carried out and who may be able to offer their advice. While it is useful to discuss other people's experiences, it is still important to carry out your own research and to minimise the risk of being disappointed by introducing some simple control methods.

Keep an open mind and be prepared to learn

It can be easy to take advice from an individual who has had a good (or bad) experience. However, agreements (and any form of undertaking by two parties) are always a two-way process. Co-operation by both parties is vital; it is not unusual for problems to occur that frustrate one or the other. Communication is the key to the successful completion of a project, and by being prepared to learn you will be in a much better position to debate on a situation that you are not clear or happy about.

Unfortunately, builders can receive a bad press and are branded as those 'rogue builders' that we often hear about in the media. At the same time, some builders work in a very professional manner and occasionally undertake work for clients who are not fully informed or as knowledgeable as they appear. Clients who are not prepared to learn and understand the facts can be the cause of their own misfortune and those of the builder.

What's new?

Home information packs

The way in which we buy houses will soon change. By the year 2007, all properties will be required to have a 'home information pack'. The UK Government plans to introduce the home information pack throughout England and Wales from the beginning of 2007. Estate agents, etc., will be required to provide copies to potential buyers on request.

Why are the home information packs being introduced?

- To ensure that buyers and sellers of properties are better prepared and have the correct information right from the start, as far as is reasonably practicable.
- To secure faster mortgage offers and replies to searches.
- To minimise delays and uncertainties.
- To help address problems, including 'gazumping' and 'chains', where buyers and sellers have been delayed.

Research has shown that, by international standards, the system currently in use is slow and inefficient, resulting in high rates of failed transactions. The home information pack will become compulsory so that buyers and sellers will benefit. Sellers will not have to wait for legislation in order to take advantage of home information packs. Many estate agents are already offering such packs on a voluntary basis.

Computers and technology will be the key to ensuring that information can be obtained and exchanged quickly and economically, although this is only in support of the home information packs and not a substitute. The physical (hard copy) home information packs are still required to be available at the very start of the transaction. It will not be a criminal offence to fail to provide a home information pack, though Trading Standards Officers will be given discretionary powers to deal with enforcement. This could be in the form of assistance, a warning or a civil fixed-penalty notice. In addition to a fixed-penalty notice, a person who breaches home information pack obligations may be sued by a prospective buyer for recovery of the costs for obtaining the appropriate documentation which should have been provided in the pack.

Research has shown that a typical transaction takes approximately eight weeks from offer acceptance to exchange of contracts. This is considered far too long and can cause problems. The new system and rules could reduce this period significantly by ensuring that the information required by the buyers and sellers is available at the start of the process. It is anticipated that the timescales for acceptance to exchange of contracts will be halved. It is important to note that the reduction of transaction-threatening risks should give both parties much more confidence that the transaction will proceed to completion.

At the time of going to press, it is estimated that the cost of preparing a home information pack for an average home will be in the region of £650. However, these are not considered additional costs as the changes to the system will require the same work to be conducted as in a survey, but they will be done more efficiently and earlier in the process. The home information pack will transfer the responsibility from the buyer to the seller for the production of a home condition report, and for obtaining local searches. However, since most sellers are also buyers, the costs would be balanced by corresponding savings and benefits. There are a number of voluntary home information pack schemes currently operating, which means that sellers may not have to pay for the pack in advance.

The benefits of the home information pack will be to bring together important information at the very start of the buying/selling process, which at present is collected during the days and weeks after an offer has been accepted. The changes will assist in:

- helping the seller to decide on a realistic asking price
- providing the buyer with the appropriate information required in order for a well-informed offer to be placed
- reducing the risk of terms being re-negotiated due to late disclosure of key information
- reducing the period between offer acceptance and exchange of contracts, thereby reducing the risk of 'gazumping' or other problems.

According to research, a home information pack could be put together in approximately ten working days, and with the use of information technology and experience, this could be even quicker. In the case of information not being immediately available to complete the pack, it will still be possible to put the property on the market while it is being gathered.

The person responsible for ensuring that the home information pack is available will be the person marketing the property. The requirement for a home information pack will apply to properties on the open market, and will not apply to:

- non-residential property
- mixed commercial (or industrial) and residential property
- properties sold with sitting tenants.

Home Condition Reports

Home information packs will include a Home Condition Report that will need to be prepared by a qualified building surveyor or similar professional who has the skill and experience to carry out an accurate inspection.

Undertaking Home Condition Inspections to required standards involves the following work to be conducted:

- Inspecting the property to ascertain its true condition.
- Recording all of the findings.
- Interpreting the evidence and determining condition ratings.
- Gathering appropriate information for establishing energy efficiency.

Preparing Home Condition Reports involves:

- preparing comprehensive reports
- making the reports available and maintaining/updating them.

Preparing for Home Condition Inspections requires inspectors to:

- identify and agree the client requirements
- make preliminary enquiries about the property
- work in an effective and professional manner
- develop and maintain good working relationships
- contribute to the safety and security of clients and their property
- safeguard the security of information.

What's in the home information pack?

The home information pack includes:

- Terms of sale
- Evidence of title
- Standard searches
- Planning consents and building control certificates
- Property information form
- Warranties and guarantee
- Home Condition Report and Energy Efficiency Assessment.

Providing a Home Condition Report

Research suggests that approximately 45 per cent of failed transactions result from problems uncovered after terms have been agreed. These problems may arise following a valuation inspection or home condition report/survey. At the time of going to press, the cost to consumers can be around £1,000 per transaction. Even if the transaction goes ahead, considerable delays can occur while re-negotiations take place. Where there is a chain of properties awaiting the outcome of one situation, the effects can go well beyond those directly involved.

The Home Condition Report provides considerable advantages for both sellers and buyers.

- It helps sellers to make decisions on the asking price for their home.
- It highlights work that needs to be carried out.
- It gives the seller the opportunity to have the work done before putting the property on the market.
- Quotes for the cost of the work (that could be passed on to the buyer) can be obtained.
- Buyers can make an offer for the property based on the facts.
- The risk that the transaction fails or is delayed due to problems with the condition being revealed at a later stage is reduced.

The Home Condition Report is a report on the condition of the property that can be reasonably relied upon by buyer, seller and lender. The report covers matters of importance to a buyer. For example:

- The general condition of the property taking into account:
- Age
- Character
- Location
- Energy efficiency
- Defects or other matters requiring attention.

The Royal Institute of Chartered Surveyors is currently working on a new format designed to give prospective buyers the correct information. The home condition report will be similar to the current homebuyers' survey and valuation, but without the valuation element. Because of the fluctuation in property prices, valuations will be left out of the survey. The property will need to be valued at the time it goes on the market, and not when the home condition report is carried out. Only inspectors/surveyors who are qualified under a certification scheme approved by the Secretary of State will be permitted to prepare home condition reports. The scheme will incorporate systems and procedures for monitoring and auditing inspectors' work to ensure that the prescribed standards are maintained.

The home condition report will provide a 'snapshot' of the condition of the property at the time it was inspected, and as it does not include a valuation, it should be reliable for some time. It is generally considered that decisions on whether the report needs to be updated should be left to buyers, sellers and their professional advisers. The length of time between the report being assembled and other circumstances (such as additional work being carried out to the property) would determine the details of any decisions to update it.

New homes will not be required to have a home condition report, provided that they are being sold with a recognised warranty. However, any subsequent sale will make a home condition report necessary even if it is within the warranty period. Final decisions on these and many other issues surrounding the home condition report are currently still under consideration.

The inspection

The inspection itself will take considerably longer than the Mortgage Valuation Inspection that lending institutions currently require. It is expected that a typical three-bedroom, two-reception house of around 90–100 square metres will take anything from 75 minutes to two hours to inspect. Some other work may be required to research aspects of the property. However, this would possibly be carried out at the inspector's office. The reports will be generated electronically and compatible with buyers', sellers', lenders' and other professionals' technology. Apart from providing a general summary of the property, the report is designed to highlight urgent or serious defects.

The report is divided into a number of sections – A–H:

- **Section A** describes the extent of the inspection and terms of engagement of the Home Inspector. It also draws attention to parts of the property that the report does not cover. The report has 'condition ratings'; Section A describes what each rating means.

- **Section B** provides a general summary of information about the property. The information covered in this section is used by lending institutions where buyers are taking out a mortgage. It includes information about the type of accommodation, reinstatement cost for insurance purposes, and it highlights those parts of the property where there are defects, which generates more information later in the document.

- **Section C** draws attention to health & safety points that are not 'defects', and provides for information that will need to be collected from the conveyancer and others.

- **Section D** lists elements of the property that are inspected from the outside, for example the roof covering, brickwork, rendering, windows, doors, etc. Each element is given a 'Condition Rating' with a statement on whether remedial action is required and whether it is urgent or serious.

- **Section E** lists all elements inspected from inside the property, for example roof structure, ceilings, floors, and walls, together with main fittings such as kitchen and bathroom. A sub-section also allows for information about damp. As with the exterior inspection, all elements are given a 'Condition Rating'.

- **Section F** provides for the visual inspection of parts of the services that can be seen, such as electrical components, gas facilities, water, heating and drainage. A 'Condition Rating' is given to each part. This is a visual inspection only and no tests are carried out.

- **Section G** reports on all of the permanent outbuildings, such as garages, sheds and conservatories; also garden walls, retaining walls and paved areas. Common facilities (such as communal areas and facilities for blocks of flats) are also reported here. No 'Condition Ratings' are awarded in this section.

- **Section H** delivers the energy efficiency report. It sets out how energy-efficient the home is, and (if appropriate) what can be done to make it more so.

☻ The final section of the report is for the Home Inspector who undertook the inspection to certify it.

Provide peace of mind

In this book I have tried to provide you with the necessary information to plan and run a project, either by using individual tradesmen, contractors or builders. It is important to understand that if you do not plan your project properly in advance, you will have problems in implementing procedures that you decide to introduce as the project gets under way. Your project may only require certain elements covered in this book, alternatively there may be issues that are not covered. Since the construction industry and each individual project is diverse, you will need to implement procedures and systems that suit the degree of difficulty or complexity of your project.

Only by setting out a plan of action and studying it will you be able to manage your project with confidence. A competent and co-operative builder, contractor or tradesman will respect the efforts that you make to familiarise yourself with the facts and knowledge for the project. Once you have demonstrated that you are prepared to learn and can implement site-management procedures, you will have provided yourself and others with 'peace of mind'.

Appendix

sample specification

The information needed to compile a specification would be taken from the main drawings and would include the details that are required to comply with the planning permission and building regulations. The architect and engineer would normally include building regulation information as a bare minimum on the drawings and/or in a separate document.

All other information in the specification is a detailed description of the client's required finishes, types of material to be used, and may include information regarding the level of standards required. From the client's point of view it is important to see evidence that the builder or contractor can produce work to the standards required.

The following is a sample based on the type of specification that a professional surveyor or Contract Administrator may prepare. A professional specification would include reference to specific legal clauses. It would identify the materials much more comprehensively including manufacturer's names and contact details. Architects and surveyors have databases that they can tap into to call up the information and they are, of course, familiar with the products required.

As detailed in Chapter 3, the specification is in four sections as follows:

- Section One   Preliminaries and general conditions
- Section Two   Specification/scope of works
- Section Three   Summary and form of tender
- Section Four   Appendices

although only Sections One and Two are reiterated below.

Section One: Preliminaries and general conditions

Section 1A – Project particulars

- Nature of work
  - Whether your project is a new build, extension or refurbishment it needs to be described for clarification of the project. For example:
    - construction of detached four-bedroom house
    - refurbishment of semi-detached three-bedroom house
    - two-storey extension and loft conversion.
- Address
  - Although the address will be on all drawings it is important that it is referred to in all contractual documents even if the address for correspondence is different from the project address.
- Timescale
  - Specified number of weeks or best programme commencing from a preferred start date, which would be subject to all documentation and approvals being secured and agreed in writing.
- Client's name
  - All contractual documents need to include the client's full name.
- Name and address of Contract Administrator, if appropriate
  - Where a Contract Administrator has been appointed to act on behalf of a client, in any capacity for all or part of the project, their full name and contact details need to be included.
- Name and address of Quantity Surveyor (if appropriate)
  - Where a Quantity Surveyor has been appointed to act on behalf of a client, in any capacity for all or part of the project, their full name and contact details need to be included.
- Name and address of Engineer, if appropriate
  - Where an Engineer has provided structural drawings for the project, their full name and contact details need to be included.
- List of all Tender and Contract Documents
  - Site plans
  - Elevation drawings
  - Electrical drawings
  - Mechanical drawings
  - Drainage plans
  - Main drawings 1–22
  - Structural drawings 1–15
  - Sections A–A ~ B–B ~ C–C ~ D–D

Section 1B – The site and existing buildings

- Details of existing buildings on the site
  - Description of buildings. For example:
    - Detached three-bedroom house with brick-build shed.
- Existing mains or services (normally part of appendices)
  - Drawings as supplied by statutory undertakers.
  - Care to be taken to ensure utility services that run through the site are protected.
- Site investigation reports
  - If a site report has been prepared this would normally become part of the tender documents or attached in the appendices.
- Access to site via specified route
  - It may be appropriate to specify a preferred route particularly if there are special circumstances prevailing such as private roads or restrictions from other routes. For example:
    - Site vehicles to avoid approaching site from Eastern Road.
- Parking and any restrictions
  - Parking of all contractors' and sub-contractors' vehicles is restricted solely to the site.
  - Parking in the street directly outside the site is prohibited.
- Surrounding land and building uses
  - The surrounding area is residential and full co-operation is required to ensure consideration for neighbours, with minimum disturbance, is a priority.
  - Where noisy work is unavoidable, method statements will be required to demonstrate the control measures that will be put in place.
- Risks to health & safety
  - The soffits at eaves level are known to contain levels of asbestos. Correct procedures and appropriate documentation will be required to remove and dispose of the material.
  - Lead pipes are known to exist in the original property; appropriate PPE (personal protective equipment) must be used when handling this and any other hazardous material.
- Site visits
  - Before providing a detailed tender, the contractor is to visit the site in order to be familiar with the nature of the project and to assess any potential problems with regard to access.
  - All local conditions and restrictions should be fully understood.
  - Arrangements for visiting site can be made by contacting …

Section 1C – Description of the work

- The work
  - Demolition of existing three-bedroom detached house with all existing foundations removed.
  - The existing property currently has gas, electric, water and sewer services, all of which enter the property below ground.
  - Telephone cables are currently servicing the property from a telegraph pole positioned in the street within 10m of the property.
  - The proposed building work consists of the construction of a four-bedroom detached house on piled foundations.
- Completion by others
  - Elements to be fully organised by client:
    - all floor finishes
    - all ceramic tiling work
    - supply and installation of kitchen
    - provision of light fittings
    - provision of sanitary ware.
- Shrinkage and cracks
  - The contractor will be expected to return to the property on expiration of the agreed defects liability period to make good all shrinkage cracks and gaps that have occurred in the new work.
  - All costs for remedial work associated with shrinkage will be borne by the contractor.

Section 1D – Contract of Agreement

- Type of contract
  - There are contracts available off the shelf, which are designed to meet the needs of homeowners who are employing the services of builders or contractors. These are usually used on the larger projects.
  - A formal contract needs to be drawn up for any project no matter how small (see Chapter 3).
- Earliest date of possession
  - It is anticipated that work will start on or around —/—/— subject to confirmation.
- Date of completion
  - The work is anticipated to take approximately 24 weeks; however, precedence over acceptance of tender may be given to proposed programme dates by contractors.
- Extension of time
  - Extension of time may be negotiated subject to:
    - additional work by instruction
    - delays in related work being organised by client

- adverse weather conditions
- design changes by client resulting in delays
- delays caused by unforeseen circumstances other than those specified (subject to discussion).

- Penalty for late completion
  - At the time of going to press, a penalty of £400 per week or part of a week will be imposed for late completion of the project, subject to any extension of time being agreed.
- Defects liability period
  - The period of defects liability will be six months from Practical Completion.
- Valuations
  - The first valuation will take place 21 days after the project start date with following valuations being carried out on the same day in four-weekly intervals thereafter.
- Payment dates
  - Payment will be made by cheque within 14 days of agreed valuations.
- Insurance details
  - All relevant copies of insurance policies must accompany the tender.
- Disputes
  - In the event of a dispute an independent arbitrator will be appointed who is a member of the Royal Institute of Chartered Surveyors and experienced in dispute resolution.

Section 1E – Tendering

- Scope of works and preliminaries
  - Only works as stated in the preliminaries and scope of works should be tendered for.
- Exclusions
  - If the contractor is unable to tender for any of the work as stated in the preliminaries or scope of works, the client or Contract Administrator (CA) must be informed as soon as possible.
  - Details of the work that has not been tendered for and the reasons for exclusion must be made clear.
- Acceptance of tender
  - The client does not guarantee that the sender of the lowest tender will automatically be invited to enter into a contract.
  - The client will not be liable for any costs in the preparation of tenders.
- Period of validity
  - Tenders must remain valid (unless withdrawn) for a period of at least 14 weeks from the date of submission.
- Pricing of specification and clarification
  - It is important to invite those tendering to seek clarification on any issues that they may not fully understand.

- All works in the scope of works and shown on the drawings should be detailed sufficiently to cover all works, other than the exclusions noted above.
- The priced specification/scope of works
  - All elements that are individually described should be individually priced.
  - Where individual elements are not priced, it will be deemed that they have been included elsewhere in the tender.
  - Where qualification or alterations to the specification need to be made for tendering purposes, they will need to be verified by the client or CA prior to submission of tender.
- Errors in the priced specification
  - If it is found that there are errors in the drawings or specification, the builder must inform the client as soon as it becomes apparent.
  - Corrections must not be made without notifying the client or CA.
- Programme of works
  - A proposed programme of works will be required with the tender.
  - If there are periods required for planning or designing these should be clearly identified.
  - Timescales for work that is being arranged by the client should be included in the programme.
- Substituted products or material
  - If the contractor proposes to use different products from those which have been specified by the client, full details must be included in the tender for verification by client or CA.
  - Materials substituted at construction phase may not be accepted if they were not brought to the attention of the client at tender stage.
- Quality control procedures
  - A statement detailing the resources for controlling the quality of workmanship and materials must accompany the tender, including:
    - material compliance procedures
    - quality control of sub-contractors' work
    - number and type of staff
    - qualifications of supervisory staff and their duties.
- Health & safety information
  - The following documents must be submitted with the tender:
    - a statement detailing how the contractor proposes to undertake health & safety obligations to safeguard all site operatives, sub-contractors, visitors and all persons who may be affected by the work
    - a copy of the contractor's health & safety policy
    - general risk-assessment procedures
    - details of persons responsible for health & safety for the project with details of their qualifications.

Section 1F – Documents, definitions and interpretations

- ◐ Qualification of wording (generally)
  - CLIENT: the owner or person for whom the project is being carried out
  - CONTRACT ADMINISTRATOR (CA): the person or persons who has/have been nominated to act on behalf of the client and who has/have the authority to discharge specific actions
  - APPROVAL or VERIFICATION: the approval/verification in writing to or from the client or CA
  - SUBMIT: documents to the client or CA unless stated otherwise
  - PRODUCTS: materials and goods intended to be permanently incorporated into the work
  - REMOVE: disconnect and/or dismantle as necessary and remove from site (other than materials or equipment owned by statutory undertakers)
- ◐ In the case of a project which involves refurbishment or alterations the following terms may apply:
  - KEEP FOR REUSE:
    - prevent damage occurring to the stated material or component (so far as is reasonably practicable) and clean off any surplus jointing or bedding material
    - store in a safe place on-site or as indicated by the client or CA, re-use or incorporate as indicated
  - REPLACE:
    - remove the stated materials or components as indicated and replace with materials that are equivalent in quality and visual aspects
    - make good any areas that have been disturbed in order to carry out the replacement work
  - REPAIR: carry out remedial work to material, component, features or finishes as indicated, to leave in a condition that matches the original finished condition
  - MAKE GOOD: carry out remedial work to areas that have been disturbed by work carried out under this contract
  - EASE:
    - to make appropriate adjustments to moving parts in order to achieve a standard of work that enables the good working operation of the stated component
    - ensure good fit and free movement; make good as required
  - TO MATCH EXISTING:
    - to use materials or products that match as closely as possible the visual characteristics and features of the existing work
    - to produce work that when finished is as inconspicuous as possible in physical and visual appearance

- Additional documentation
  - Two copies of all current drawings and specification will be issued free of charge to the contractor prior to submission of tender.
  - Two copies of all further revised drawings and contract documentation will be issued during the contract free of charge.
  - Additional copies of drawings and other documentation will be issued on request and charged per item:
    - Architect's drawings £...
    - Engineer's drawings £...
    - Detailed drawings £...
    - Specification £...
    - Other documentation £...
- Dimensions
  - Dimensions scaled from drawings are not guaranteed and it will be the contractor's responsibility to produce work to accommodate pre-made or manufactured components.
  - Where discrepancies or ambiguity relating to specific dimensions occur, clarification must be obtained from the client or CA before proceeding.
  - Where appropriate, the contractor will be responsible for producing the correct dimensions for work associated with building regulations and planning permission criteria.
- Documents and drawings provided by contractor/sub-contractors
  - Where drawings are required for fabrication or construction purposes (other than those supplied by the client) two copies must be issued to the client or CA for approval. It is the contractor's responsibility for meeting relevant building control or planning permission requirements.
- Technical literature
  - The contractor must keep on-site all necessary technical information for supervisory personnel and trades persons to refer to as and when required, for example:
    - manufacturer's literature
    - relevant BS Codes of Practice
    - performance ratings, where appropriate.
- Maintenance instructions and guarantees
  - Contractor to retain on-site all documentation related to any material, product or component.
  - Contractor to register guarantees with the appropriate manufacturer on installation of product, and deliver copies to client or CA.
  - Contractor to provide the client or CA with telephone numbers of sub-contractors responsible for emergency repair services.

Section 1G – Management of the works generally

- Supervision
  - Contractor to arrange for supervision, co-ordination, and general administration of all works including that of all sub-contractors.
  - Contractor to ensure continuity of work to each trade by arranging and monitoring the programme accordingly and taking into account local authority inspections, statutory utilities work and delivery of material.
  - All supervisory staff to be experienced, trained and qualified to carry out their duties.
- Weather conditions – keeping records
  - Accurate records must be kept of daily minimum and maximum air temperatures.
  - Good building practices must be used with regard to working in extreme weather conditions.
  - Work affected by adverse weather conditions should be recorded to include:
    - description of weather conditions
    - dates
    - type of work affected
    - total number of hours lost
    - damage caused as a consequence.
- Ownership of material
  - All redundant material arising from demolition or alteration becomes the property of the contractor. It must be removed from site at the earliest opportunity, or as work proceeds.
  - Only material stated in the specification for retaining or materials owned by statutory utilities are to be kept, as indicated or fixed.
- Programme and progress
  - Before starting the work on-site a master programme must be submitted for approval by the client or CA and must include:
    - Design and production of information by the appropriate persons who are required to do so for construction purposes, including checking and inspections
    - Planning and mobilisation by the main contractor
    - Carrying out the main scope of works.
  - Where the work is being organised by the client the contractor must seek advice from the client or CA.
  - A progress report detailing progress against programme must be prepared and submitted to the client or CA two days prior to site meetings.
  - A record of progress must be kept on-site and updated on a daily basis.
  - Proposals for recovering lost time due to unforeseen circumstances must be put forward as and when requested.

- Meetings
  - A pre-contract meeting will take place between the client and contractor one week prior to commencement of project.
  - Regular fortnightly meetings will take place on dates to be arranged.
  - The contractor will be expected to hold meetings with sub-contractors to ensure standards and continuity are maintained.
- Notice of completion
  - In order for the client's independent inspections to take place, at least one week's notice must be given prior to the whole or parts of the project reaching practical completion.
- Adverse weather – protection
  - The contractor will be expected to use all reasonable methods to prevent or minimise delays to the project during adverse weather conditions.
  - The contractor will also be expected to minimise any potential damage that could be caused by adverse weather conditions by employing the use of suitable protection methods.
- Extension of time
  - Where the contractor intends to apply for an extension of time, the following information must be provided at least two weeks before the commencement date of such extension of time:
    - reason for extension of time
    - effects of the delay on other work
    - revised programme showing new completion date.
  - Extension of time will only be granted if evidence for reasonable justification is produced.
- Cash flow forecast
  - As soon as possible and in conjunction with the valuation dates, the contractor must produce a forecast of payments required for completed works as indicated on the main programme.

Section 1H – Instructions and information flow

- Proposed instructions from client
  - Where the client or CA issues details of a proposed instruction with a request for estimated costs, the contractor must submit the estimate as soon as possible and in any case within seven days to include:
    - a detailed breakdown of cost for labour and material or as indicated
    - details of additional resources required
    - details of adjustments to the programme.
  - If the contractor is unable to submit an estimate for a proposed instruction, the client or CA must be informed immediately.
- Required information
  - The contractor or sub-contractors who require information must provide a formal written request, to include:

- details of the information required
- dates by which the information is required.
- The client or CA must be given at least three days to provide the appropriate information.
- Where the client or CA cannot provide information within three days due to third-party input, the contractor will be informed.
- The client or CA will endeavour to provide information or answers to questions as soon as is reasonably practicable.

- Measuring completed work
  - The contractor is required to give reasonable notice for work that is not immediately visible in order for inspections and measurements to be carried out.
  - Any work that has been carried out in excess of or below that stated in the specification/scope of works needs to be brought to the attention of the client or CA as soon as it becomes apparent.

- 'Day work' agreements
  - Where a day work agreement has been negotiated, reasonable notice must be given prior to the start of day work taking place, and a record of day work must be kept to include:
    - reference of instruction for the nature of day work
    - name of person(s) carrying out the work
    - materials used
    - plant used
    - hours spent on the work
    - signature of person in charge.

- Interim valuations
  - Information on work carried out must be submitted to the client or CA five days prior to the established valuation dates with full details of the amounts due in relation to work carried out.
  - The valuation must include no more than:
    - fixed materials provided by the contractor
    - labour cost of fixed materials
    - percentage of preliminary costs in proportion to elapsed project time
    - a list of unfixed material on-site
    - a list of unfixed material or products off-site.

- Unfixed material on-site
  - The client or CA retains the right to pay for a percentage of unfixed material on-site.

- Listed off-site material or goods
  - The client or CA retains the right to pay for a percentage of materials or goods as listed off-site.
  - Confirmation of payment or deposits paid for listed materials off-site will be required.

Section 1J – Quality control and standards

- ⊖ Good practice
  - Where details of materials, products and workmanship are not fully specified, they are required to be of a standard in keeping with British standards and suitable for the function for which they are intended.
  - The contractor, sub-contractors and individuals will be expected to work within the scope of good building practice and as a minimum meet the following conditions:
    - have consideration for others
    - keep work areas clean and tidy
    - work in a safe manner
    - be environmentally friendly
    - be responsible for their actions
    - be accountable for their actions.
- ⊖ Material – compliance
  - All materials must be appropriate and suitable for the purpose for which they are being used, and must comply with the appropriate British or European standards.
  - Care should be taken to ensure that materials and products comply with the specification and Building Control requirements.
  - Specific checks should be made to ensure that:
    - quantities of delivered materials are correct
    - finishes and colours are correct and match approved samples
    - sizes and dimensions of materials or products are as specified on drawings or technical information
    - materials with a specified shelf life are not out of date
    - materials are delivered clean and undamaged.
- ⊖ Consistency
  - Where large orders of similar materials are expected to be delivered over a period of time, the contractor must make certain that methods are used to ensure colour matching or correct batching processes are implemented.
- ⊖ Protection of material
  - The contractor must ensure that all practical measures are taken during the construction phase to ensure that materials or products are fully protected whether they are fixed, unfixed, boxed, wrapped, etc., until practical completion has taken place.
  - All original packing cases must be retained until practical completion has taken place.
- ⊖ Workmanship
  - The contractor must ensure that operatives are experienced and skilled in the type of work they are expected to carry out.

- All work that involves products or elements with moving parts must operate freely and without binding or grinding.
- Samples of material for approval
  - Where specified the contractor will be responsible for providing samples of a product or material for approval by the client or CA.
- Samples of finished work
  - Samples of the finished work will be required for approval:
    - Brickwork 1.0m x 1.0m panel
    - Plastering 1.5m x 1.5m
    - Painting
- Setting out – discrepancies
  - Contractor to check and record the levels of the site in relation to those indicated on the drawing.
  - Dimensions to be checked for accuracy and any discrepancies in levels or dimensions to be reported to the client or CA before proceeding with work.
- Appearance and fit
  - Contractor to be responsible for all setting out and construction of all building elements and finishes (working in conjunction with the drawings, design and specification) to ensure correct fit and appearance.
  - Where satisfactory appearance or accuracy of fit are difficult to achieve, approval of proposals must be obtained from the client or CA as soon as possible.
- Tolerances
  - Contractor to work in accordance with the British Standard (BS 5606) for tolerances in construction.
- General services
  - All work affecting utility services must be in accordance with by-laws, regulations and statutory authority requirements.
- Mechanical and electrical tests
  - All M + E works to be carried out in accordance with the regulatory bodies' codes of practice.
  - All M + E work to be inspected/tested during construction and a final test carried out on practical completion.
  - All test and installation certificates to be handed over to the client as and when appropriate.
  - Statements that the work complies with the regulations for gas and electrical installation will be required.

Section 1K – Inspection and co-ordination of work

- Supervision
  - The contractor will be responsible/accountable for the supervision and actions of all site operatives including sub-contractors and visitors to site.

- Co-ordination of building and engineering services
  - The contractor will be responsible/accountable for all work including sub-contract work.
  - At least one person on-site under the direction of the main contractor should have sufficient knowledge of the work and project to ensure cohesion between all trades.
- Access for inspection by client or CA
  - Access facilities for the client or CA to inspect work must be made available by the contractor.
  - At least one week's notice must be given before removing access facilities such as scaffolding.
- Timing of tests and inspections
  - Sufficient notice must be given where inspections or tests need to be witnessed by client, CA or Engineers.
  - Dates and times should be made for inspections where possible.
- Test certificates
  - Copies of certificates for any tests carried out should be submitted to the client or CA, and a copy kept on-site where appropriate.
- Proposal for rectifying defects
  - Where defects have been found, written proposals for rectifying the work should be forwarded to the client or CA for approval.
- Quality control and test records
  - A system to ensure that all works are carried out in accordance with the specification should be put into operation, for example:
    - Keeping a record of inspections or tests to include:
      - element of work, date and location
      - nature of the inspection
      - result of the inspection/test
      - details of corrective work required.
- On completion of work – making good
  - Contractor to make good at their expense any damage caused to the client's property, neighbouring property, footpath or road during the construction work.

Section 1L – Security ~ Safety ~ Protection

- HSE approved codes of practice
  - Contractors to comply with H & S Law.
  - If the health & safety of site personnel, visitors or any persons affected by the work is put at risk in the opinion of the client or CA, the work will be stopped until the contractor provides written documentation of how this will be addressed.

- Security measures
  - The contractor will be expected to make adequate arrangements for the security of:
    - the works
    - material
    - plant
    - access to adjoining properties.
- Stability
  - The contractor will be responsible for the stability and structural integrity of the works during the contract.
  - Method statements and calculations for support work may be required; contractor to liaise with engineer.
- Occupied premises
  - Additional care and health & safety measures will need to be taken in occupied premises.
  - Proposals for safeguarding others should be submitted to the client or CA.
  - Where work is required to be carried out during periods when the client is in occupation, the client will bear the cost of additional work such as:
    - moving and protecting furniture
    - temporary protection work
    - working outside of normal working hours.
- Noise
  - The contractor must comply with (BS 5228-1) noise control on construction sites.
  - The contractor must make every effort to minimise the effect of noise during the work, and report any problems associated with noise to the client or CA.
- Pollution
  - Adequate measures must be taken to control pollution to the following:
    - water supply
    - air
    - soil.
  - If contamination does occur, the client or CA must be informed without delay of all details.
- General nuisance
  - All precautions must be taken to reduce the risk of general nuisance, for example:
    - noise
    - smoke
    - dust
    - vermin
    - light
    - fumes.

- Asbestos-based materials
  - If asbestos is found during the project (other than that stated) it should be reported immediately to the client or CA.
- Fire prevention
  - The contractor must take all necessary precautions to prevent personal injury to site personnel, or damage to the property as a result of fire.
  - Smoking will not be permitted on-site and all appropriate fire-fighting equipment must be available and adjacent to any work being carried out that poses a risk.
  - A fire plan must be available on-site and all site operatives must be aware of the procedures in the event of a fire.
  - As part of the tender a proposal for fire prevention must be submitted.
- Burning on-site
  - Burning of redundant material or waste will not be permitted on-site.
- Moisture
  - The contractor must make all necessary arrangements to protect materials and completed works from the effects of excessive moisture content.
  - Drying out procedures must be carried carefully to avoid:
    - blistering
    - trapped moisture
    - cracking or excessive movement.
- Infected timber
  - Where infected timber is required to be removed, this must be done in such a way as to minimise the risk of affecting other parts of the building.
- Waste
  - All waste or surplus material must be regularly cleared from site and the site kept clean and tidy.
  - All waste, including hazardous waste, must be disposed of at the appropriate tips as approved by the waste regulation authority.
  - Documentation of all waste transfers must be retained for inspection.
- Existing services above or below ground
  - Contact all service providers to obtain the appropriate drawings and locations of services.
  - Adequately protect and prevent damage to existing services, and do not interfere with them.
  - Where required, contact the appropriate service provider for advice on the correct recommendation for working close to their supplies.
- Roads and footpaths
  - Maintain public routes for roads and footpaths adjacent to the site.
  - Keep roads and footpaths clear of mud, debris and materials at all times.
  - Any damage to roads or footpaths that may result from the building work. must be made good to the satisfaction of the local authority or owner.
  - Contractor to pay for any damage caused.

- Existing topsoil/subsoil/trees/shrubs
  - The contractor will be responsible for repair and costs in relation to damage caused to surrounding elements in general.
- Adjoining property
  - Arrangements for access to adjoining properties for erecting scaffold or carrying out essential work should be made via the client or CA.
  - Any damage caused to adjoining properties will be made good as directed by the client or CA.
  - The contractor will bear the cost of any remedial work required.
- Existing structures
  - Methods of excavations or demolition of work adjacent to adjoining property or existing structures must be prepared prior to commencement.
  - All appropriate methods of support, and shoring, strutting, etc., must be employed to ensure structural integrity during the work.

Section 1M – Facilities and services

- Locations – site facilities plan
  - A plan must be provided with the tender to indicate the position of:
    - soil heaps
    - site office/meeting room
    - welfare facilities
    - toilet
    - material storage
    - skip
    - temporary water supply
    - temporary electrical supply
    - position of service runs.
- Meetings
  - Details of arrangements for holding meetings must be provided with tender.
- Welfare/sanitary accommodation
  - Clean and maintained sanitary accommodation must always be made available for client and site operatives.
- Lighting
  - In order for the client or CA to inspect finished work appropriate lighting must be provided when required.
- Water – restrictions
  - Details of water provisions and emergency arrangements must be provided with tender.
  - In the event of emergency water facilities being required, this will be paid for by the client.
- Telephone
  - The contractor must provide, as soon as possible after starting on-site, a direct telephone line to the person in charge.

- ⊖ Fax
  - ■ The contractor must provide, as soon as possible after starting on-site, a fax line or facility.
- ⊖ Email
  - ■ Email facilities are desirable for the contract, but not essential.
  - ■ In the event that email facilities are not provided, the minimum requirement will be a fax line.
- ⊖ Meter readings
  - ■ Meter readings will be taken at possession and completion of the project, and costs of usage will be paid for by the contractor.

Section Two: Specification/scope of works

Note: All sections of the specification/scope of works must be read in conjunction with the main contract, preliminaries and general conditions.

Section 2A – General information

- ⊖ This section may include specific information that, although touched on in previous sections, may need more clarification, for example:
  - ■ a fuller description of the overall work
  - ■ that health & safety information such as the company H & S policy together with Risk Assessments and Method Statements will be required for specific elements of work
  - ■ that the contractor will be responsible for costs incurred due to damage caused by delivery lorries or other associated parties
  - ■ that scaffolding work needs to have a separate programme for erection duration and dismantling, and all measures to protect the neighbouring properties from falling debris must be taken
  - ■ details of what the programme should include
  - ■ the priced tender/specification should show each item individually priced as it appears on the summary page
  - ■ that the highest standard of work will be expected, and where the client is not satisfied the contractor will reproduce the work at their own expense
  - ■ that all material and labour will be deemed to have been allowed for within the price, except where specifically noted by the client for exclusion.

Section 2B – External works/demolition

- ⊖ Prior to demolition work, make arrangements for an asbestos identification survey to be carried out. This should include a report for safe removal to a licensed disposal site. Allow for removal and carting away – provisional sum.

- Prior to demolition, liaise with the appropriate utilities to locate and isolate all services, for example:
  - gas
  - electricity
  - water
  - communication networks.
- Submit to client (and local authority building control) method statements for the safe demolition of existing buildings.
- Protect trees, surrounding fences and brick walls identified on drawing No....
- Erect temporary hoarding to secure the site during the demolition stage.
- Carefully remove and retain for re-use existing ornamental fireplace.
- After safe removal of asbestos-based material, carefully demolish and remove existing building materials, including:
  - roof covering and structure
  - windows, doors and frames
  - floor joists
  - all brickwork
  - foundations and slab.
- As indicated on drawing, remove trees, shrubbery and foliage, excavate and remove roots.

Section 2C – Excavations and foundations

- Excavate to reduced level as shown on drawing, and put to one side any clean topsoil for re-use by client. All other debris to be removed from site.
- Allow for the protection of all services, for example:
  - drain runs
  - gas mains
  - water mains
  - electrical services.
- Carry out piling as per engineer's drawing to depths shown and report to client if depths are exceeded with the additional costs on a pro rata basis.
- Construct ground beams in accordance with engineer's drawings.

Section 2D – Brickwork, block work and all associated work

- As indicated on drawing No... construct cavity wall from ground beam up to damp-proof course level using class B engineering bricks.
- Damp-proof course to be bitumen-based class A and lapped 110mm where necessary with adhesive as per manufacturer's recommendations.
- Provide sample of face bricks as specified, including a built sample of 1.0m x 1.0m for approval prior to main construction work.
- Construct main walls with a weathered struck mortar joint, and 75mm cavity-

filled insulation as identified on drawing No.... Inner leaf of cavity walls to be constructed of 100mm standard blocks as identified on drawing No....

- Allow for expansion joints where indicated on drawing No....
- Build in stainless steel vertical twist type wall ties at appropriate dimensions, for example:
  - 900mm centres horizontally and 450mm vertically
  - 225mm centres vertically at all openings and within 150mm of opening.
- Allow for all openings as indicated on drawings with lintels as specified on lintel schedule, minimum 150 bearing.
- Cavity trays and weep holes in accordance with manufacturer's recommendations.
- Where indicated on drawing fix telescopic air vents at 1.800 centres.
- Cavity closers type ... to be installed as per drawings at all openings.
- Spray weed killer to the entire internal area of the new building.
- Construct 200mm sleeper walls in accordance with engineer's drawings, built off main ground beams.

Section 2E – Suspended block and beam ground floor and associated work

- Construct block and beam floor as per drawing, manufacture and design to be arranged by contractor and details submitted to client's engineer for approval prior to construction.

Section 2F – Scaffolding

- Allow for appropriate scaffolding to enable the safe construction of the proposed building works.
- Keep all pathways and entrances clear of scaffold at all times.
- All necessary security measures to be taken to provide security when site is not manned, for example access ladders removed and locked up at the end of each working day.
- Any damage caused by the erecting or dismantling of scaffold will be rectified at the contractor's own expense.
- All scaffolding work must comply with the appropriate health & safety regulations.

Section 2G – Roof structure, covering, fascia, soffits and bargeboards

- As indicated on drawing No... construct roof with roof trusses as specified, including wall plates and all bracing. All strapping to be in accordance with details on drawing No....
- Roof insulation to be fixed to the pitched section between rafters and over main ceiling joists/rafters as per drawing No....
- Roof tiles to be concrete plain tiles as specified on drawing No... 268mm x

165mm fixed on 38mm x 25mm pre-treated timber battens with high-performance non-tear roof felt lapped into gutter.

- Dry vent ridge system to be fixed to manufacturer's recommendations.
- Eaves vents to be installed into soffit.
- Fascia, soffits and bargeboards to be PVC-u system as shown on drawing No....

Section 2H – Rainwater goods

- As indicated on drawing No... 112mm half round PVC-u gutters and 68mm round down pipes including all necessary fittings, colour to be black and manufactured by ....
- All fixings in accordance with manufacturer's recommendations.
- Terminate into gullies at ground level in positions as indicated on drainage drawing No....

Section 2J – External windows and doors

- Supply and fix PVC-u windows and doors manufactured by ... type ..., all windows to be factory finished with lockable handles and fasteners.
- Trickle vents and 4/16/4 double-glazed units with low E glass. Restrictors to be fixed to accessible casements. Obscure glass type ... for bathrooms and toilets.
- Toughened glass to be used for windows below 800mm.
- Door to front (DG-01) type ... with three sets of keys.
- Door to rear (DG-07) type ... with three sets of keys.
- Window boards to be compatible and fixed to manufacturer's recommendations.
- A ten-year guarantee for windows, doors and all components will be required on completion and forwarded to the client.
- Window and door installation must be carried out by a FENSA- (Fenestration Self Assessment) registered company.
- Certification of installation will be required on completion.

Section 2K – External drainage

- In accordance with drawing No... excavate to required depth for inverts in order to construct manholes 1 and 2 on a 150mm concrete base.
- Manholes to be semi-engineering bricks (English bond) bedded in 1:3 (sand–cement).
- Bench all work and point brickwork in accordance with good working practice
- Excavate and bed in back inlet gullies (surrounded in C25 concrete) as indicated on drawing No....
- Excavate and lay pipe work type ... laid in accordance with building regulations requirements.
- Terminate new drainage into existing manhole 3 as indicated on drawing No....
- Fully test drainage runs on completion and on practical completion.

Section 2L – External paving and driveway

- As indicated on drawing No... excavate and allow for new block paving to be made up of:
  - 100mm MOT type 1 base
  - 50mm well-compacted sharp sand
  - 200mm x 100mm block paving type ....
- Ensure that all falls are correct to displace water to gullies as indicated on drawing No....

Section 2M – Internal walls, floors and door frames

- First floor joists to be constructed as indicated on drawing No....
- Joists to be strapped to wall at 1.200mm centres.
- Joists to be fixed at 200mm centres below each bath tub position.
- Solid strutting to be fixed at 2.0m centres.
- Fix end block to last joist and wall.
- Where partitions are specified allow to double up joists and bolt at 400mm centres with M12 bolts to include double side toothed plate connectors between joists.
- All walls on ground floor to be built of 100mm lightweight blocks and taken through floor joists.
- Pre-cast concrete lintels to be 100mm wide x 65mm deep reinforced with 150 bearing each end over doorways.
- All walls to ground floor to be restrained at high level with 30mm x 5mm straps at 2.0m centres.
- All walls to first floor to be 100mm x 50mm sawn softwood, pre-treated.
- All studs to be at 400mm centres with two rows of noggins equally spaced.
- Infill studs with 75mm insulation as indicated on drawing No....
- Lintels to be 150mm deep and let into double studs at doorways.
- Allow for radiator noggins in positions shown on drawing.
- Allow for half-hour fire-resisting door frames at all door openings manufactured by ....

Section 2N – Internal doors and ironmongery

- All doors to be supplied by ...; for type of door see schedule.
- Ironmongery to be supplied by ...; see door schedule.
- All hinges to be 75mm stainless steel double washer type.
- The door schedule would not normally be shown in the specification, but would be a separate document as an appendix.

INTERNAL DOOR AND IRONMONGERY SCHEDULE						
DOOR	TYPE	HINGES	HANDLES	LEVER LOCK	LEVER ONLY	FLOOR STOP
DG-02	PF/02	3	Ref. 1234	No	Yes	No
DG-03	PF/02	3	Ref. 1234	No	Yes	Yes
DG-04	PF/02	3	Ref. 1234	No	Yes	Yes
DG-05	PF/40	3	Ref. 1234	Yes	No	Yes
DG-06	PF/40	3	Ref. 1234	Yes	No	No
DF-01	PF/02	3	Ref. 1234	No	Yes	No
DF-02	PF/02	3	Ref. 1234	No	Yes	No
DF-03	PF/02	3	Ref. 1234	No	Yes	No
DF-04	PF/02	3	Ref. 1234	No	Yes	Yes
DF-05	PF/50	3	Ref. 1234	Yes	No	Yes
DF-06	PF40	3	Ref. 1234	Yes	No	Yes
DF-07	PF40	3	Ref. 1234	Yes	No	No

Section 2P – Kitchen units and fittings

- Kitchen to be supplied and fitted by specialist contractor arranged by client.
- Building contractor to allow for supply of services only to positions as indicated on drawing No....

Section 2Q – Staircase and associated works

- Staircase and associated works all to be in softwood and constructed as indicated on drawing ...; supplier of all stair material to be ....

Section 2R – Plasterboard, plastering and coving

- Fix 12.5mm plasterboard to all ceilings and stud walls with 3mm skim finish.
- All internal plastering to be 11mm thick with 2mm skim finish coat including galvanised beading on all corners.
- All rooms to have S profile cornice size ....

Section 2S – Plumbing, heating and sanitary fittings

- All plumbing and heating to be installed as per drawing No....
- All hot and cold pipe work to be copper, and sizes as indicated on drawing.
- All wastes to be polypropylene, and sizes as indicated on drawing.
- Radiators to be type ... manufactured by ....
- Sanitary ware to be provided by client.
- Boiler to be sized sufficiently to allow the possible addition of two more radiators.

Section 2T – Electrical installation and fittings

- All electrical work should be carried out in strict accordance with the current I.E.E (Institute of Electrical Engineers) Wiring Regulations.
- Carry out electrical work as indicated on drawing No....

- All switch and socket faceplates to be chrome type ....
- Light fittings to be provided by client.
- Dimmer switches to all rooms except cupboards and loft.
- Aerial and telephone points to all rooms except bathrooms.
- Smoke detectors to be type ....
- All cables to be chased into wall and protected with conduit.
- Install burglar alarm as indicated on drawing No..., manufacturer ....
- The following heights apply to the centre of faceplates from finished floor level:
  - consumer unit 1.800mm
  - lamp holder 2.100mm
  - light switches 1.200mm
  - all low-level sockets 450mm
  - cooker switch 250mm (above worktop)
  - sockets in kitchen and utility 250mm (above worktop)
  - room thermostat 1.200mm
  - telephone, TV and data points 450mm.

Section 2U – Floors and finishes

- Lay 65mm screed to ground floor as per detail drawing No... including mesh and insulation.
- Lay 18mm chipboard flooring to all first floor areas allowing for moisture-resistant in bathrooms and en-suite bathrooms.
- Client to arrange supply and fitting of all floor finishes.
- Building contractor to allow for:
  - 9mm WPB plywood to en-suite bathrooms and shower room
  - 9m^2 of 18mm chipboard flooring to loft area – storage.

Section 2V – Ceramic wall tiling and mirrors

- Client to arrange the supply and fitting of all ceramic tiles.
- Client to arrange supply and fitting of all mirrors.

Section 2W – Internal decorations

- Prepare, seal and paint with the following:
  - Ceilings and cornices all rooms. Paint to be manufactured by ...
    - 1 No. mist coat and two coats white vinyl matt emulsion.
    - Allow for vinyl silk to bathrooms, en-suite bathrooms, kitchen and utility room.
  - Walls – all areas. Paint to be manufactured by ....
    - 1 No. mist coat and two coats vinyl matt emulsion, client to advise on colours during project.
    - Allow for vinyl silk to bathrooms, en-suite bathrooms, kitchen and utility room.

- Joinery work, all areas.
  - rub down
  - apply knotting
  - prime
  - apply filler where appropriate
  - 1 No. undercoat
  - 2 No. high gloss (white)

glossary

1st fix carpentry The fixing of floor joists, roof timbers, partitioning, stairs and door liners

2nd fix carpentry The fixing of skirting, architraves, doors and door furniture, kitchen units, wardrobes, etc.

Acrows Temporary structural supports used primarily in refurbishment work

Architrave Timber moulding fixed around outer edge of door frame

Asbestos Mineral composed of thin, flexible fibres, commonly found in and on older properties, which is very hazardous to health

Bar chart Graph used to illustrate such as timescales for specific elements of work

Barge board Fascia covering to the apex of a roof

Battery drill Rechargeable battery-powered drill

Bitumen Tar-like compound commonly used as an adhesive for flat felted roofs

Block paving Man-made concrete blocks of various sizes, colours and shapes commonly used for driveways, paths and patios

Block work Blocks laid to form the main structure of the house, usually used for the inner skin and internal walls

Bracing Timber or metal braces usually fixed at angles to provide rigidity to the structure to which it is fixed

C25 concrete Concrete produced to a specific design requirement, e.g. C25 is used in foundations

Cavity closers Commonly a plastic insulated barrier placed between the inner and outer skin of brick and block work prior to installation of windows

Cavity trays A membrane laid over lintels and between the inner and outer skin of block work to deflect moisture through the weepholes

Cavity wall Two walls forming the structure of a building usually with a cavity space of approximately 75mm, filled with insulating material

Celcon lightweight blocks Proprietary brand of lightweight block

Chipboard interlocking boards Commonly used as floorboards within domestic properties

Chopsaw Specialised saw used in construction for cutting timber

Cladding Thin covering or overlay, e.g. of stone on a building or metal on a metal core

Compact dumper Motorised vehicle used for transporting earth on building sites

Compressor Mechanical or electrical device used to produce compressed air for the operation of pneumatic tools

Consolidated hardcore Compacted rubble used as a sub-base

Cornice Decorative moulding used at the abutment of ceilings and walls

Damp-proof course Membrane laid between brickwork courses at ground-floor level, usually 150mm above external ground level, to prevent rising damp

Data cabling Cabling used for such as TV, computer, telephones, alarms, etc.

Decking Decorative garden structure commonly made of wood

Door liners Timber frame fixed into structural door opening onto which door is hung

Dormer window A window set vertically in a structure projecting through a sloping roof

Double side tooth plate connector Structural component used to prevent two timbers from moving

Drainage gulley Pre-formed clay or plastic unit into which surface water or waste water flows

Dry lining Plasterboard used to line walls instead of plaster or render

Dry vent ridge system System used within roofing to ventilate roof space (loft)

Eaves The lower border of a roof that overhangs the wall

Electrical breaker Miniature circuit breaker that opens in the event of excessive current, which has to be manually closed when the overloading problem has been investigated and put right

Elevation drawings A representation of the front, back or side of a building

Faceplates Word commonly used to describe light switches, sockets, etc.

Fascia A flat, horizontal piece of such as stone or board under projecting eaves

Floor joists Structural timbers spanning from wall to wall onto which the floorboards are fixed

Footings The base of a structural wall

Gazumping To thwart (a would-be house purchaser) by raising the price after agreeing to sell at a certain price

Grinder Electrical tool generally used for cutting blocks, steel, or forming slots for cabling

Hardcore Crushed concrete, or rubble

Hydraulic working platform Mechanised platform for working at heights

Isolation valves Valves or stop cocks used to isolate the flow of water or gas

Jigsaw Electrical tool used for cutting out shapes

Joists (general) Timbers that form either flooring or ceilings

Kango Electrical tool used for breaking up concrete or hard surfaces

Lintel A horizontal architectural member spanning and usually carrying the load above an opening

Load-bearing wall A supporting structural wall

M12 bolt Specific diameter of a bolt, e.g. M12 = 12mm

Mastic sealant A sealant which provides a waterproof joint, e.g. window frame and wall

Membrane (breathable) For example, roofing felt which allows air to flow out of the roof structure but will not allow water to penetrate it

Membrane (polythene) For example, a membrane laid over the entire area of an extension to prevent moisture from rising

Meter head position The point at which electrical cabling enters the property

Mineral wool insulation Material used to meet the insulation qualities required for building control compliance

Mortar joint The cement joint between the horizontal and vertical joints of brick and block work

Muck away lorry Vehicle used for removing large quantities of excess earth, debris, etc.

Newel post cap Decorative cap used on the principal post supporting either end of a staircase handrail

Noggin A horizontal timber, e.g. to affix a radiator onto a timber wall

Outerskin (of brickwork) The brickwork that forms the outer surface of a cavity wall structure

Outriggers Part of a tower scaffold to provide stability

Partition wall Wall dividing two rooms

Party wall A common wall dividing two properties

PAT (testing) Portable Appliance Test (for electrical tools and equipment)

Percussion drill An electrical device for drilling into hard surfaces such as concrete, brick, etc.

Plant Term commonly used to describe machinery, e.g. dumpers, diggers, forklift trucks, concrete mixers, etc.

Plasterboard Board with a plaster core used as a substitute for plaster on interior walls, onto which plaster is usually added

Pneumatic breaker Compressed-air operated tool for breaking up surfaces such as thick concrete

Pointing Mortar or cement used to fill joints in brickwork

Point loading The transfer of weight down to a specific point

Polypropylene Any of various plastics or fibres that are polymers of propylene

Practical completion The point at which a project is virtually complete

Pricework Work that is carried out to an agreed overall cost, or on a rate per square metre

PVC-u Polyvinyl chloride unplasticised, commonly used in the manufacture of windows, doors and other building materials

Radon Radioactive gas

Reinforcement mesh Steel mesh used to reinforce concrete

Rendering Plaster applied to a wall

Retaining wall A wall used to hold back earth

Roof trusses Timber frames designed and manufactured from plan to form the shape of the roof

Sand blinding A layer of sand applied to hardcore on which polythene membrane is laid

Screed A mixture of sand and cement applied to a floor to give it a level surface

Semi-engineering bricks Non-porous brick used for foundations, drains and other wet or damp conditions below ground. Also used in structural brickwork for its strength

Septic tank A tank in which the solid matter of continuously flowing sewage is disintegrated by bacteria

Skip A large open container for waste or rubble

Skirting A board, often with decorative moulding, that is fixed along the base of an interior wall

Soffit The underside of an overhanging part of a building, staircase, arch, etc.

Spreader boards Boards used to distribute load

Strapping Commonly used to secure roof structures and timbers to brickwork

Support props Temporary supports used during construction work (e.g. excavations and refurbishment)

Telescopic air vents Vents inserted into the brick and block work to form an airflow beneath ground floors

Timber studs Upright timber framework onto which plasterboard is usually fixed

Tolerance An allowable variation from a standard dimension

Tower sections Part of a lightweight access scaffold

Trickle vent A ventilation unit commonly found in modern windows

Trip switch Mechanism which cuts off the flow of electricity in the event of overloading

Underpinning To form part of, strengthen, or replace the foundation of a building or other structure

Wall plate A plate fixed to the top of a wall which secures roof trusses

Wall tie A manufactured metal component used to tie the inner and outer skin of brickwork

Weathered struck mortar joint A type of pointing (see Pointing, above)

Weep holes Holes left in the vertical joints of brickwork to allow moisture to escape

further information

ODPM: Office of the Deputy Prime Minister

Free and priced publications concerning government legislation/guidelines can be obtained from:

Office of the Deputy Prime Minister Publications
PO Box 236, Wetherby LS23 7NB
(Tel: 0870 1226236; Fax 0870 1226237; odpm@twoten.press.net)

Schemes of registered installers

Fenestration Self-Assessment Scheme
Fensa Ltd, 44–48 Borough High Street, London SE1 1XB
(Tel: 0870 7802028; Fax: 020 7407 8307; www.fensa.co.uk)

Oil Firing Registration Scheme
OFTEC, Foxwood House, Dobbs Lane, Kesgrave, Ipswich IP5 2QQ
(Tel: 0845 6585080; Fax: 0845 6585181; www.oftec.org)

HETAS Ltd, 12 Kestrel Walk, Letchworth, Hertfordshire SG6 2TB
(Tel: 01462 634721; Fax: 01462 674329; www.hetas.co.uk)
(Registration scheme for companies and engineers involved in the installation and maintenance of domestic solid fuel-fired equipment)

The Council for Registered Gas Installers (CORGI)
1 Elmwood, Chineham Park, Crockford Lane, Basingstoke, Hampshire RG24 8WG
(Tel: 01256 372200; Fax: 01256 708144; www.corgi-gas-safety.com)

Authorised competent person self-certification schemes for installers who can do all electrical installation work
BRE Certification Ltd
Bucknalls Lane, Garston, Watford, Herts WD25 9XX
(Tel: 0870 6096093; www.partp.co.uk)

British Standards Institution
Maylands Avenue, Hemel Hempstead, Herts HP2 4SQ
(Tel: 01442 230442; www.bsi-global.com/kitemark)

ELECSA Ltd
44–48 Borough High Street, London SE1 1XB
(Tel: 0870 7490080; www.elecsa.org.uk)

NAPIT Certification Ltd
The Gardeners Lodge, Pleasey Vale Business Park, Mansfield, Notts NG19 8RL
(Tel: 0870 4441392; www.napit.org.uk)

NICEIC Certification Services Ltd
Warwick House, Houghton Hall Park, Houghton Regis, Dunstable, Beds LU5 5ZX
(Tel: 0800 0130900; www.niceic.org.uk)

Authorised competent person self-certification schemes for installers who can do electrical installation work only if it is necessary when they are carrying out other work:
- CORGI
- ELECSA Ltd
- NAPIT Certification Ltd
- NICEIC Certification Services Ltd
- OFTEC

ODPM Building Regulations Division and the Welsh Assembly Government

Office of the Deputy Prime Minister
Buildings Division, 18/B Portland House, Stag Place, London SW1E 5LP
(Tel: 020 7944 5742; br@odpm.gsi.gov.uk; www.odpm.gov.uk/building-regulations)

The Welsh Assembly Government
Construction & Domestic Energy Branch, Housing Directorate, Crown Buildings,
Cathays Park, Cardiff CF10 3NQ
(Tel: 029 2082 6922; housing.branch@wales.gsi.gov.uk; www.wales.gov.uk)

Information and literature

Approved Documents for Building Regulations

Copies of the Approved Documents are available for purchase from The Stationery
Office (TSO), (Tel: 0870 6005522; Website: www.tso.co.uk).

- Approved Document A – Structure: 2004 Edition Published by TSO, 2004 ISBN 0-11-753909-0
- Approved Document B – Fire Safety: 2000 Edition, Published by TSO, 2004 ISBN 0-11-753911-2
- Approved Document C – Site preparation and resistance to contaminants and moisture: 2004 Edition, Published by TSO, 2004 ISBN 0-11-753913-9
- Approved Document D – Toxic Substances: amended 1992, ninth impression, Published by TSO, 1999 ISBN 0-11-751821-2
- Approved Document E – Resistance to the passage of sound, 2003 Edition, Published by TSO, 2002 ISBN 0-11-753642-3
- Amendments 2004 to Approved Document E – Resistance to the passage of sound, Published by TSO, 2004 ISBN 0-11-753915-5
- Approved Document F – Ventilation: 1995 Edition amended 2000, seventh impression, Published by TSO, 2003 ISBN 0-11-752932-x
- Approved Document G – Hygiene: 1992 Edition, second impression (with amendments) 1992, further amended 2000, Published by TSO, 2003 ISBN 0-11-752443-3
- Approved Document H – Drainage and waste disposal 2002 Edition, Published by TSO, 2001 ISBN 0-11-753607-5
- Approved Document J – Combustion appliances and fuel storage systems, 2002 Edition, Published by TSO, 2001 ISBN 0-11-753494-3
- Approved Document K – Protection from falling, collision and impact, 1998 Edition, Published by TSO, 1997 ISBN 0-11-753390-4
- Approved Document L1 – Conservation of fuel and power in dwellings, 2002 Edition, third impression 2003 with revisions, Published by TSO, 2003 ISBN 0-11-753609-1
- Approved Document L2 – Conservation of fuel and power in buildings other than dwellings: 2002 Edition, second impression, Published by TSO, 2002 ISBN 0-11-753610-5
- Approved Document M – Access to and use of buildings, 2004 Edition, Published by TSO, 2003 ISBN 0-11-753901-5

- Approved Document N – Glazing – safety in relation to impact, opening and cleaning, 1998 Edition, amended 2000 seventh impression, Published by TSO, 2003 ISBN 0-11-753389-0
- Approved Document P – Electrical safety, Published by TSO, 2004 ISBN 0-11-753917-1

Other Priced Publications (available from TSO)

- Accessible thresholds in new housing: Guidance for house builders and designers. TSO March 1999 ISBN 0-11-702333-7
- Approved Document J: 2002 Edition: Guidance and Supplementary Information on the UK Implementation of European Standards for Chimneys and Flues. TSO October 2004 ISBN 0-11-753919-8
- Building Regulations and Fire Safety – Procedural Guidance. TSO February 2001 ISBN 0-11-753596-6
- Manual to the Building Regulations Third Edition 2001. TSO December 2001 ISBN 0-11-753623-7
- BS 5440: Part 1 2000. Installation and maintenance of flues and ventilation for gas appliances of rated input not exceeding 70 kW net (1st, 2nd and 3rd family gases). Specification for installation and maintenance of flues. December 2000 ISBN 0-580-33229-2
- BS 5440: Part 2 2000. Installation and maintenance of flues and ventilation for gas appliances of rated input not exceeding 70 kW net (1st, 2nd and 3rd family gases). Specification for installation and maintenance of ventilation for gas appliances. December 2000 ISBN 0-580-33098-2
- BS 5440: Part 2 2000 Amendment No 14912. July 2004
- BS 7671: 2001: Requirements for Electrical Installations IEE Wiring Regulations – 16th Edition – incorporating Amendments No 1: 2002 and No 2: 2004. IEE 2004 ISBN 0-86341-373-0
- Safety in the installation and use of gas systems and appliances. HSE November 1998 ISBN 0-717-61635-5

Free literature

Copies of these publications are available from ODPM Publications (see full contact details at the beginning of this section):

- A Guide to Determinations and Appeals (product code 01CD0751)
- Building Control Performance Standards (product code 99CD0352)
- New rules for electrical safety in the home (product code 04BR02710)
- Solid Fuel, Wood and Oil Burning Appliances. Get them checked, sweep your chimneys, and be safe (product code 99ASC0638)
- The Party Wall etc. Act 1996 Explanatory Booklet (product code 02BR00862)
- Your Garden Walls, Better to be SAFE... (product code 91HCN0227)

The above publications are subject to change. An up-to-date list can be found on the ODPM website.

- Gas appliances. Get them checked – Keep them safe (product code INDG238(REV2)
- Landlords: A guide to landlord's duties: Gas Safety (Installation and Use) Regulations 1998 (product code INDG285 (REV1)

The above leaflets are available free by mail order from HSE Books, PO Box 1999, Sudbury, Suffolk CO10 2WA (Tel: 01787 881165; Fax: 01787 313995; Website: www.hsebooks.co.uk)

- Radon – a guide for homebuyers and sellers (product code 00EP414B)
- Radon – a guide to reducing levels in your home (product code 00EP414C)
- Radon – You **can** test for it (product code 00EP0414A)
- Radon – a householder's guide (product code PB9442)
- Protocol on Design, Construction and Adoption of Sewers in England and Wales – produced in support of Approved Document H listed above (product code PB6472)

The above guides are available free from DEFRA Publications, Admail 6000, London SW1A 2XX (Tel: 08459 556000; Fax: 020 8957 5012; defra@iforcegroup.com)

- Building Regulations and Historic Buildings – Balancing the needs for energy conservation with those of building conservation: an Interim Guidance Note on the application of Part L (product code 50675)

This is available free from English Heritage Customer Services, PO Box 569, Swindon SN2 2YR (Tel: 0870 3331181; Fax: 01793 414926; Customers@english-heritage.org.uk)

- Need a plumber or builder…?
  - Part A – A step-by-step guide to getting work done on your home
  - Part B – Organisations which can help you get work done on your home
- You will be sent both parts unless you specify otherwise (product codes OFT 118a & b)

These are available free from OFT, PO Box 366, Hayes, Middlesex UB3 1XB (Tel: 0870 6060321; oft@eclogistics.co.uk; or download them from the Website: www.oft.gov.uk)

Miscellaneous

The planning portal

The portal contains extensive information about the planning system in general, whether planning permission is likely to be needed and how to go about making a planning application

(Tel: 0117 3728885; Fax: 0117 3728804; www.planningportal.gov.uk)

Practical Completion – agreeing the end of the contract with the builder

Further information and advice on construction law and contracts can be obtained from:

Alway Associates

Banbury Head Office, 3 West Bar, Banbury, Oxon OX16 9SD

(Tel: 01295 275975; Fax: 01295 275981; enquiries@alway-associates.co.uk; www.alway-associates.co.uk)

Insurance details regarding self-build

Further information covering site insurance, structural warranty and building control can be obtained from:

Self Build Zone

London House, 77 High Street, Sevenoaks, Kent TN15 1LD

(Tel: 0845 2309874; sales@selfbuildzone.com; www.selfbuildzone.com)

Literature

A comprehensive range of construction and related books, covering architecture, building surveying, quantity surveying, building services, building skills, health & safety, regulations and contracts, interior design, home improvements and self-build, is obtainable from:

The Building Centre

26 Store Street, London WC1E 7BT

(Tel: 020 7692 4040; Fax: 020 7636 3628; bookshop@buildingcentre.co.uk; www.buildingcentrebookshop.co.uk)

index

acknowledgements

I would like to thank Carey Smith and Natalie Hunt at Ebury Press for their constant help and support, and Two Associates and Jonathan Baker for their excellent design work. I wish to express my appreciation to my editor, Margaret Gilbey, who has been incredibly helpful and has willingly provided her professional expertise.

I am grateful for the information supplied by the following reference sources, without which it would have been impossible to compile this book accurately: The Chartered Institute of Building, The Office of the Deputy Prime Minister, The Livemore Partnership (TLP), The Royal Institute of Chartered Surveyors, The Health and Safety Executive, The Joint Contracts Tribunal Ltd, The Building Centre (London), Alway Associates, Mark 1 Hire, Paul Thorne Associates (Structural Engineers) and Construction Management Services 99 Ltd.

I would also like to thank Mr & Mrs Owen, Melvin Napper and all those businesses and individuals who have provided me with a wealth of invaluable information, essential to the writing of this book. You know who you are!

And finally, special thanks to: Cyndy Phillips-Hoogervorst MBE, Stan the man, Tony Watts, Pippa Harper, Ruth Walters, Roger Kemp, Clive Potter, Johnny (Kango) Waldock, Joe Cooke, Tracey Smart, and to my family. They have all given their encouragement, time and advice willingly, from the moment the idea for this book was conceived, to seeing it on the shelves.

This book is dedicated to the memory of my parents, Len and Pat Sales.

Building peace of mind

also from ebury

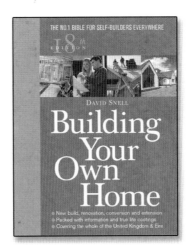

Building Your Own Home

18th Edition
By David Snell

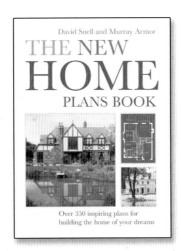

The New Home Plans Book

By Murray Armor and David Snell